The Chemistry of Uranium

nova science publishers
www.novapublishers.com

Chemistry Research and Applications

More information about this series can be found at
https://novapublishers.com/product-category/series/chemistry-research-and-applications/

Nuclear Materials and Disaster Research

More information about this series can be found at
https://novapublishers.com/product-category/series/nuclear-materials-and-disaster-research/

William A. Alexander
Editor

The Chemistry of Uranium

nova
science publishers
www.novapublishers.com

DOI: https://doi.org/10.52305/IRKT5368

Library of Congress Cataloging-in-Publication Data

ISBN: 979-8-89530-802-8 (Hardcover)
ISBN: 979-8-90134-009-7 (eBook)

Published by Nova Science Publishers, Inc. † New York

Contents

Preface

This volume contains chapters covering various aspects of the Chemistry of Uranium. It is not comprehensive and could not attempt to be so. Rather than attempt to include the whole breadth of possible topics, each chapter has been authored in a way to take a deep dive into a particular area of Uranium Chemistry. Within Section A, "Engineering Information," Charatsidou et al., present an introduction to uranium production and use as a nuclear fuel. The chapter is written with a keen eye toward the future of the nuclear power industry, with discussion of waste management and next-generation reactor technology. Palanki has produced an authoritative coverage of uranium materials processing, focusing on UO_2 processing and sintering for the fuel pellets that are the primary source for current generation of nuclear reactors. The amount of knowledge contained within that chapter is impressive and reflects the author's decades of experience.

In Section B, "Energetic Interactions," Anderson, Charbonnet, and I present chapters on aspects of UF_6 chemistry, detailing reactions between UF_6 and organic surfaces, and a computational study investigating UF_6 interactions with a host of organic functional groups.

In Section C, "Environmental Implications," Patni and a team of co-authors focus on uranium content in water bodies from a global perspective, with a particular examination of these impacts in India. Finally, work from the Cerrato group at the University of New Mexico investigating uranium's environmental fate and transport properties from mine wastes is summarized.

I thank all the contributors for their participation and patience while this volume was assembled.

Also, a bit of fun. While I was in graduate school, the university student paper started printing daily sudoku puzzles. Some of my peers and I would race each other to see who could complete them the quickest. They became an occasional bit of good-natured fun and distraction between experiments and simulations. As a bit of fun, this volume provides interspersed sudoku puzzles between the chapters. They are all themed around the Chemistry of Uranium, provided by the puzzle setter *rockratzero*. The first is a "classic" sudoku, with subsequent puzzles being from the universe of "variant sudoku." These variant sudoku puzzles have lines, dots, and arrows included within the grid to provide extra rules and constraints to the number placement. Rules are given for each puzzle, and a QR code and URL are provided to point the reader to an online solving platform as well. Once one wraps their head around the rulesets, the puzzles present nice logical workouts in a different way than "classic" sudoku. The later puzzles are a bit tricky, with the final one lacking any given digits! Do give them a go!

William A. Alexander, Editor

Section A

Engineering Information

With a fourth generation of nuclear power, you can have a technology that will burn more than 99 percent of the energy in the fuel. It would mean that you don't need to mine uranium for the next thousand years.

- James Hansen

Puzzle 1

Classic Uranium

by rockratzero

9								2
	1	6				4	9	
	2	7				5	1	
	7	2				6	5	
	3	4				9	2	
	9	5	6		7	3	8	
		1	5	6	9	8		
			2	3	8			

Rules

Place the digits 1-9 in each row, column, and 3x3 box without repeats.

Link to SudokuPad online: https://tinyurl.com/rrz-uranium1.

Chapter 1

An Introduction to Uranium Use, Waste Management, and Advanced Reactors

Elina Charatsidou[1,*]
Pär Olsson[1]
and William A. Alexander[2,†]
[1]Nuclear Science and Engineering, Department of Physics,
KTH Royal Institute of Technology, Stockholm, Sweden
[2] Department of Chemistry, The University of Memphis, Memphis, TN, USA

Abstract

Uranium is the key element in the nuclear industry, with fission providing energy for electricity production but also neutrons for significant advancements in medicine, scientific research, and other diverse fields. This chapter provides a brief overview of the current state of uranium use within and around the context of nuclear power generation. The challenges and opportunities associated with uranium use are highlighted; these encompass crucial areas such as nuclear security measures, comprehensive waste management strategies to ensure environmental safety for future generations, and ongoing technological advancements that continue to shape this dynamic landscape.

* Corresponding Author's Email: elinach@kth.se
† Corresponding Author's Email: w.alexander@memphis.edu

In: The Chemistry of Uranium
Editor: William A. Alexander
ISBN: 979-8-89530-802-8

1. Current State of Uranium Use in Nuclear Energy

In the nuclear energy sector, uranium is the primary fuel used in fission reactions in nuclear reactors. Fission can take place in heavy isotopes, primarily uranium-235, but also notably plutonium-239, when such a nucleus absorbs a free neutron and transforms to an unstable isotope that rapidly splits (fissions) into two lighter elements and a few free neutrons. This highly energetic reaction is the core physical process that releases heat that can be transformed into electricity. Most nuclear reactors operate with water as the working fluid to drive steam turbines. The water also acts as a moderator for the free neutrons that are released from the fission process. At the time these neutrons are released, they are too highly energetic to likely cause fission in the surrounding uranium nuclei. The neutrons are slowed down by the water, mostly by collisions with the hydrogen atom protons of the water molecules, to more optimal energy levels to cause further fission. The only fissile isotope (i.e., that can be fissioned by neutrons) found in nature is uranium-235. The natural abundance of uranium-235 is quite low, only 0.72%, and the most common isotope is uranium-238, which is instead called fertile, since it can be transformed into fissile plutonium-239 by neutron capture and beta decay.

The current state of uranium use in nuclear energy is characterized by its widespread availability and the existence of well-established supply chains. However, challenges such as the limited availability of high-grade uranium ores, concerns about proliferation of fissile materials, the environmental impact from mining, taking proper care of the front-end waste, and the need for nuclear energy-using countries to develop their own final repositories for highly active nuclear waste pose significant hurdles for the nuclear energy field. These challenges necessitate the exploration of alternative fuel cycles and advanced reactor designs to ensure the sustainability and security of nuclear energy.

The process of converting uranium ore to a usable form for nuclear reactors begins with the extraction and milling of uranium ore. Once the uranium ore is extracted and milled, it goes through a series of chemical processes to convert it into a suitable form for reactor use. One of the key steps in this process is the conversion of uranium ore into uranium hexafluoride (UF_6) for enrichment of the fissile uranium-235. The enriched UF_6 is further processed to produce uranium dioxide, which serves as the primary fuel source in the vast majority of nuclear power plants currently operating. During the enrichment process, uranium-235 is increased in concentration. This enrichment process that enhances the concentration of uranium-235 is crucial for producing fuel with the desired concentration of fissile material to achieve reactor criticality. The fuel in most reactors is made out of pellets of uranium dioxide (UO_2) that are stacked vertically inside long zirconium alloy fuel rods. These rods are held together in fuel bundles that are placed next to one another to configure the reactor to an appropriate state where criticality can be optimally controlled.

The byproducts of the uranium enrichment process include the uranium hexafluoride that is depleted of fissile material, referred to as depleted uranium hexafluoride (DUF6), containing primarily the isotope uranium-238. These byproducts pose certain challenges for the nuclear industry, as DUF6 is primarily considered waste. There are ongoing discussions and research regarding the possible reuse of spent fuel mixed with depleted uranium in other reactor types as next-generation fuel, apart from the current practice in some countries of using some of the depleted uranium to make mixed oxide fuel (MOX) by mixing depleted uranium with plutonium that has been extracted from the spent fuel. These discussions focus on the potential benefits of reusing spent fuel and depleted uranium, such as reducing waste and maximizing the use of available resources.

2. From Ore to Enrichment: The Transformation of Uranium

Historically, uranium extraction was performed using conventional mining and milling processes, typically through underground shafts or shallow open pits. Once extracted from the earth, the ore is transported to a mill where it is crushed to a fine powder and the uranium is leached from this powder using acid, alkaline, or peroxide solutions [1]. Most cases use sulfuric acid as the leaching agent. The product after drying and filtering of the leached concentrated uranium is referred to as yellowcake.

However, modern uranium extraction is currently dominated by the in-situ recovery (ISR) process (also called in-situ leaching or solution mining), which forgoes traditional shaft/pit mining [2, 3]. The ISR process involves pumping a lixiviant solution into the subsurface. Typical lixiviants for uranium extraction contain an oxidant such as oxygen and/or hydrogen peroxide mixed with sodium carbonate or carbon dioxide. The lixiviant dissolves the uranium in place, and is pumped back to the surface, where this solution is further processed to give yellowcake.

Modern extraction processes are able to produce a more pure product which is actually mostly brown or black, but “yellowcake” remains a common term for the final milling product. The olive green-black colored triuranium octoxide U_3O_8 is one of the most kinetically and thermodynamically stable forms of uranium and is the "standard product" yellowcake used today.

Worldwide, over 58% of uranium production is via the ISR method [3], while that number is as high as 90% annually in the United States. 2022 data from the US Energy Information Administration (USEIA) reports indicate that approximately 88 tonnes of U_3O_8 resulted from underground mining, while ~3700 tonnes of U_3O_8 was produced from ISR, with a further 6800 tonnes of potential capacity reported for ISR plants on standby [4]. However, the United States is far behind Kazakastan (42%), Canada (15%), Namibia (13%), Australia (9%), and Russia (6%) as world leaders of U_3O_8 production (totaling ~85% of world production). This is not surprising as identified recoverable uranium resources in the United States are only ~1% of the global picture, with five countries Australia (28%), Kazakhstan (13%), Canada (10%), Russia (8%), and Namibia (8%) controlling over 2/3 of the known recoverable resources [3].

Regardless of the extraction process, the yellowcake is sent to uranium conversion facilities for further processing [5]. The uranium oxide yellowcake is processed and reacted with fluorine to produce uranium hexafluoride. One typical industrial process is the uranyl nitrate process. The U_3O_8 is converted to hydrated $UO_2(NO_3)_2$ using nitric acid. This is followed by calcination in oxygen to form UO_3, which is then reduced to UO_2 using hydrogen. Uranium dioxide can also be produced by heating hydrated uranium peroxide [$UO_4 \cdot 2H_2O$)] or ammonium (or sodium) diuranate [$(NH_4)_2U_2O_7$] through calcination (See Figure 1).

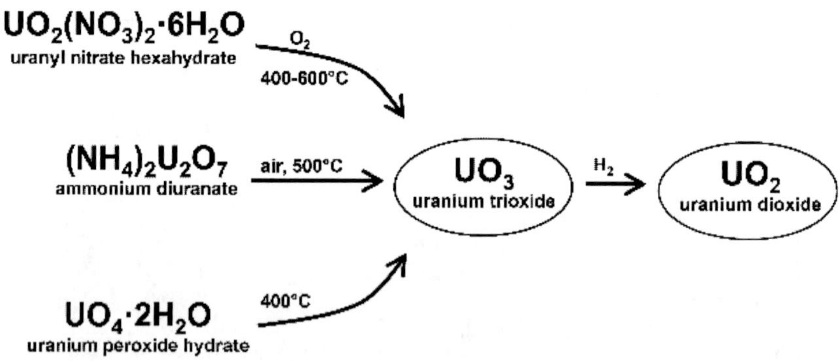

Figure 1. Possible conversion processes to obtain uranium trioxide.

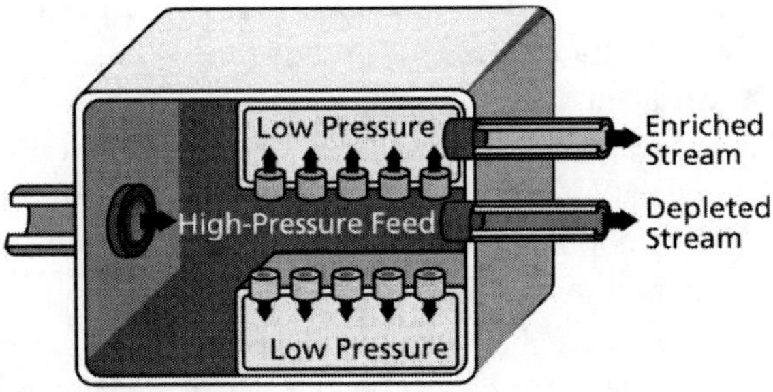

Figure 2. Schematic of the gaseous diffusion process uses molecular diffusion to separate a gas from a two-gas mixture. Isotopic separation is accomplished by diffusing UF_6 gas through a porous membrane (barrier), leveraging the different molecular velocities of the two isotopes to achieve separation. Image courtesy of the U.S. NRC [7].

The UO_2 is converted with hydrofluoric acid (HF) to UF_4 and the oxidation with fluorine gas finally yields UF_6 [6].

$$UO_2 + 4HF \rightleftharpoons UF_4 + 2H_2O$$

$$UF_4 + F_2 \rightleftharpoons UF_6$$

The very exothermic conversion of UF_4 to UF_6 is performed in a flame reactor where temperatures of about 1100°C can be reached.

The next step of the process to produce nuclear fuel involves the separation of the two major isotopes: ^{235}U and ^{238}U. While a number of enrichment processes have been demonstrated historically or at a laboratory scale, only two processes have operated at the commercial scale [7]. Both gas diffusion (aka thermal diffusion) and gas centrifugation use UF_6 for separations. The gas centrifuge process was first demonstrated in the 1940s but was not pursued in favor of the simpler diffusion process [8]. It was then developed and brought on stream in the 1960s as the second-generation enrichment technology. The economics of the centrifugation process are better at smaller scales, which enables staged development of larger plants. It is also much more energy efficient than diffusion.

Commercial uranium enrichment was first carried out by the diffusion process in the United States at Oak Ridge, Tennessee [9]. Two other plants were operated near Portsmouth, OH and Paducah, KY, with Paducah being the final plant to shut down in 2013. The three large plants were originally developed to provide high level enrichment products. It has been estimated that 7% of the total electricity demand in the United States was from these diffusion enrichment plants at the height of the Cold War, when 90% U-235 was required for the weapons program, rather than the reactor grades of 3-4% needed for power generation [9]. The diffusion method (see Figure 2) involves forcing uranium hexafluoride gas under pressure through a series of porous membranes or diaphragms. As U-235 molecules are lighter than the U-238 molecules they move faster and have a slightly better chance of passing through the pores in the membrane. The UF_6 which diffuses through the membrane is thus slightly enriched, while the gas which did not pass through is depleted in U-235. This process is repeated many times in a series of diffusion stages called a cascade. Each stage consists of a compressor, a diffuser and a heat exchanger to remove the heat of compression. The enriched UF_6 product is withdrawn from one end of the cascade and the depleted UF_6 is removed at the other end. The gas must be processed through around 1400 diffusion stages to obtain a product with a 3-4% U-235 concentration. Diffusion plants typically have a small amount of separation through one stage (hence the large number of stages) but are capable of handling large volumes of gas. Due mainly to the large physical scale and high energy requirements, the gaseous diffusion process is no longer used in the nuclear industry for uranium enrichment.

All currently operating enrichment facilities use the centrifuge process (see Figure 3). Like the diffusion process, the centrifuge process uses UF_6 gas as its feed and makes use of the slight difference in mass between U-235 and U-238. The gas is fed into a series of vacuum tubes, each containing a rotor typically 3 to 5 meters tall and 20 cm in diameter. [USEC's (now Centrus) American centrifuges are more than 12 m tall and 40-50 cm in diameter, while Russian centrifuges are generally less than one meter tall] [9]. When the rotors are spun rapidly, at 50,000 to 70,000 rpm, the heavier molecules with U-238 increase in concentration towards the cylinder's outer edge with a corresponding increase in concentration of U-235 molecules near the center. Countercurrent flow set up by a thermal gradient enables enriched product to be drawn off axially, with heavier molecules at one end and lighter ones at the other. The enriched gas forms part of the feed for the next stages while the depleted UF_6 gas goes back to the previous stage. Eventually enriched and depleted uranium are drawn from the cascade at the desired concentrations. To obtain efficient separation of the two isotopes, centrifuges rotate at very high speeds, with the outer wall of the spinning cylinder moving at between 400 and 500 meters per second, yielding a million times the acceleration of gravity. Although the volume capacity of a single centrifuge is much smaller than that of a single diffusion stage, its capability to separate isotopes is much greater. Centrifuge stages normally consist of a large number of centrifuges in parallel. Such stages are then arranged in a cascade similarly to those for diffusion. In the centrifuge process, however, the number of required stages may only be 10 to 20, instead of a thousand or more for diffusion. Centrifuges are designed to run for about 25 years continuously and cannot simply be slowed or shut down and restarted according to demand.

An important potential new technology is laser enrichment, but this technology has not yet seen commercial implementation. The principle behind laser enrichment is that laser irradiation can photoexcite specific uranium isotopes. Increasing the energy in those species alters their properties, which can be utilized to allow separation.

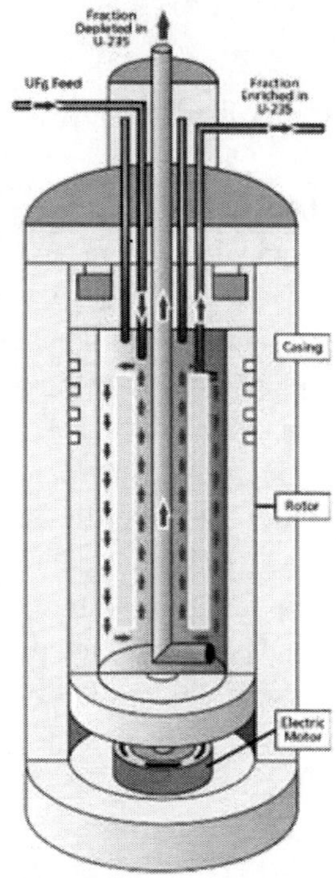

Figure 3. Schematic of the gas centrifuge process. Gas is introduced and rotated at high speed, concentrating the component of higher molecular weight toward the outer wall of the cylinder and the lower molecular weight component toward the center. The enriched and the depleted gases are removed by scoops. Image courtesy of the U.S. NRC [7].

In general, the enrichment process entails using three major components: the laser system, optical system, and separation module system. Tunable lasers can be developed to deliver a highly monochromatic light (light of a single-color). The light from these lasers can photo-ionize a specific isotopic species while not affecting other isotopic species. The affected species is then chemically changed, which enables the material to be separated. Some laser separation technologies can use a uranium metal alloy as the feed material, while the Separation of Isotopes by Laser Excitation (SILEX) method uses UF_6 as the feed material [7].

Isotope separation is crucial in the enrichment process, as it allows for the production of enriched uranium hexafluoride with a higher content of the desired isotope, ^{235}U. The enriched uranium hexafluoride is then processed further to create uranium dioxide fuel. This fuel is the main component used in the bulk of currently operating nuclear reactors for generating electricity.

3. Uranium Oxide Fuel: Production and Application

Once the enriched uranium hexafluoride has been processed, it is converted into uranium dioxide (UO_2) powder. There are mainly three methods for conversion of UF_6 into UO_2. The integrated dry route (IDR), the ammonium di-uranate route (ADU) and the ammonium uranyl carbonate route (AUC) [5, 10]. The IDR method employs hydrolysis of hexafluoride to uranyl fluoride (UO_2F_2), followed by reduction of uranyl fluoride with hydrogen at 500°C into UO_2. IDR provides several advantages such as high reproducibility of the product powder, high sinterability, and no criticality concerns since there is no water involved in the process. ADU uses hydrolysis of hexafluoride with liquid water followed by precipitation of ammonium di-uranate $(NH_4)_2U_2O_7$ and reduction to UO_2. The ADU method includes several issues associated with the conversion, such as lower reproducibility, criticality concerns due to the liquid process, as well as ammonia containing effluents. Lastly, during the AUC method, uranium hexafluoride is reacted with a solution of ammonium hydroxide and ammonium carbonate followed by reduction into UO_2 at 650°C. The advantage of the AUC method over the IDR and ADU is the production of free flowing UO_2 powder with a uniform particle size. This means that it can easily be pressed to green pellets without the use of a binder. A slight oxidation is necessary to adjust its sintering behavior and oxidation stability. UO_2 powder is pyrophoric and oxidizes spontaneously converting into U_3O_8. The controlled oxidation forms a passivating thin layer of U_3O_8 on the UO_2 powder particle surfaces to avoid spontaneous oxidation [11]. The powders produced by the three different methods will result in different powder parameters, such as specific surface area, density, morphology, O/U ratio, U content, particle size, and impurity levels.

A schematic of the nuclear fuel fabrication can be seen in Figure 4. Once UO_2 powder is produced, the next step is the fuel pellet fabrication. After the pellet manufacturing process is complete, scrap UO_2 pellets that did not pass the inspection, as well as residuals from the grinding process, are oxidized to U_3O_8 and incorporated into the UO_2 powder mix. The U_3O_8 is mixed into UO_2, as a recycling byproduct, to both minimize costs, but also to decrease pellet density and increase the mechanical strength of the green pellets. The powder mixture is blended into a homogeneous powder [12].

Following blending the powder undergoes granulation, passing through a coarse mesh and fracturing into granules. After granulation the powder is pressed into pellet shapes with the help of a die and a uniaxial press. The pressed pellet is called a green pellet. Green pellets usually have a density of around 70% of the theoretical maximum density of the pellet. The green pellets are sintered to increase their mechanical integrity and density and to achieve a solid ceramic form. The final density achieved is around 95% of the theoretical density. After the pellet has been sintered it undergoes grinding to produce uniform cylindrical pellets that meet the dimension specifications. As a final step the pellet is inspected and if the fabrication criteria are met, the pellet is moved to the fuel rod process, if not, the pellet is rejected and moved to scrap which is reused in the blending process of a new powder mixture. Quality controls are performed regarding visual aspects, dimensions, density, composition, micro- and macrostructure, and thermal stability. More technical details about each of these processing steps can be found in Chapter 2 of this volume.

These fuel rods are the primary component for nuclear energy production. They are composed of uranium dioxide fuel pellets and encased in a cladding material, usually made of zirconium alloy. Next, the fuel rods are sealed to isolate the fuel pellets from the coolant and to prevent the release of radioactive materials. The rods are then bundled into fuel assemblies which comprise the nuclear reactor core.

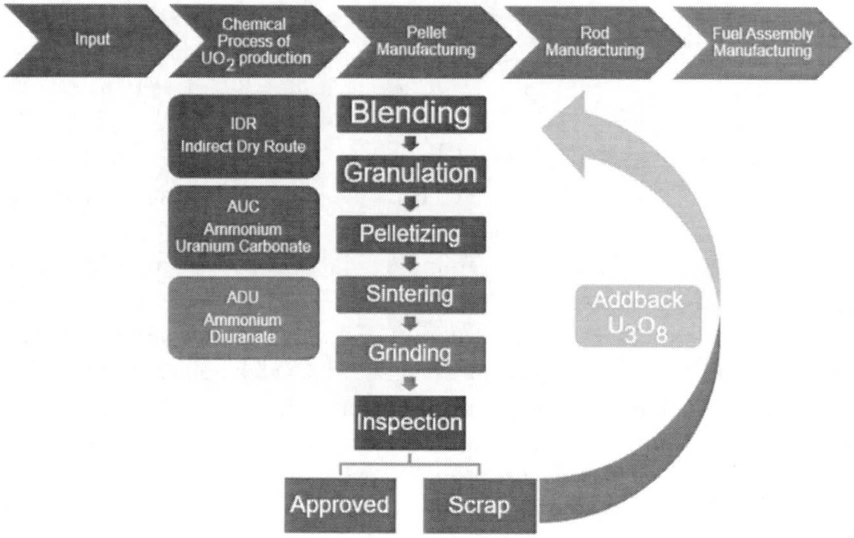

Figure 4. Schematic of the nuclear fuel fabrication cycle.

4. Byproducts of Uranium Enrichment and Reactor Operation

The process of enriching uranium, from manufacturing nuclear fuel to nuclear reactor operation, generates several byproducts, each with its own set of management challenges. While spent fuel from reactor operation often garners the most attention due to its high radioactivity, depleted uranium hexafluoride, the leftover material after uranium enrichment, presents a different set of challenges. Depleted uranium hexafluoride contains a lower concentration of the fissile isotope, uranium-235, compared to natural uranium. Currently, large quantities of DUF6 are stored on-site in sealed containers, as it is primarily considered a byproduct without a large-scale use.

Unlike spent fuel, where protection from radioactive emissions is paramount, the primary safety concerns with DUF6 revolve around its reactive and corrosive hexafluoride chemistry. UF_6 is hygroscopic (meaning it readily absorbs moisture from the air) and reacts quickly with water to form uranyl fluoride (UO_2F_2), releasing hydrofluoric acid in the process. When heated, UF_6 can also decompose, releasing dangerous fluoride fumes. While the radiological risks are lower compared to spent fuel, the sheer volume of DUF6 generated far exceeds that of spent fuel, posing significant storage and management challenges.

The physical properties of UF_6, particularly its behavior at different temperatures and pressures, necessitate careful handling procedures. Although used in gaseous form during enrichment, UF_6 is typically stored and transported as a solid. However, UF_6 undergoes significant volume changes with temperature variations. The thermal volume expansion of UF_6 is among the largest known. For example, when transitioning from a solid to a liquid, UF_6 can expand its volume by over 30%. This makes temperature control crucial throughout the handling process.

Furthermore, for UF_6 to be handled as a liquid, it needs to be kept under specific pressure and temperature conditions. If these conditions are not maintained, the UF_6 may rapidly convert to gas, potentially leading to leaks or ruptures. So, while not as radioactive as spent fuel, depleted UF_6 presents unique challenges due to its chemical properties and the large volumes generated.

On the other hand, we have spent nuclear fuel. From the nuclear reactor operation, spent fuel rods have to be taken out after 3-5 years of operation and processed in some way. The treatment of these spent fuel rods as waste is detailed in the next section. In these rods are lower levels of U-235 but a host of other radionuclides and waste materials that have been generated due to the nuclear fission. One of the radionuclides is plutonium, which is produced in small quantities during the operation of nuclear reactors. Plutonium is a radioactive element with various isotopes, some of which are fissile, making it valuable as a fuel for nuclear reactors and in nuclear weapons.

Another byproduct of nuclear reactor operation consists of transuranic elements, such as neptunium, americium, and curium. These elements are formed through the neutron capture reactions in the reactors and are of concern due to their long-lived radioactive properties. Proper management and disposal of these transuranic elements are essential to mitigate potential environmental and health risks.

Aside from radionuclides, other waste materials, such as contaminated equipment, protective clothing, and packaging materials used in nuclear facilities, also contribute to the overall radioactive waste stream. Proper management and disposal of these materials are necessary to ensure the safety of workers and the surrounding environment.

Overall, the management and disposal of byproducts of uranium enrichment and nuclear reactor operations, including radionuclides and other waste materials, require comprehensive strategies to address the complex nature of radioactive waste and minimize environmental and health risks.

5. Waste Management and Storage of Spent Fuel

Used fuel retrieved from a nuclear reactor, after 3-5 years of use, contains around 95% of uranium-238, about 1% of uranium-235 and around 1% plutonium-239 produced by neutron capture in U-238. The spent nuclear fuel emits both radiation from the fission fragments, and heat. In the short-term, the spent fuel is stored on-site for this heat and radiation to decrease to a level that is manageable for further processing.

There are essentially two options for used fuel: the so-called open fuel cycle, including long-term storage and final disposal, and the closed fuel cycle comprising reprocessing and recycling of spent nuclear fuel.

In the case of the open fuel cycle, many counties have adopted an approach for permanent disposal of spent nuclear fuel in underground geological repositories. Sweden was one of the leading counties in researching the potential of a permanent geological repository solution [13]. The main idea is that the fuel assemblies containing fuel rods and spent nuclear fuel pellets will be encapsulated in a copper canister with a cast iron insert. The canister will be positioned a few hundred meters down in the bedrock and surrounded with bentonite clay. This concept comprises six defense barriers, the fuel pellet, the fuel rod, the cast iron insert, the copper canister, the bentonite clay and lastly the bedrock itself. Extensive research was done to determine the most suitable site and a one-of-a-kind underground laboratory was constructed in Äspö near Oskarshamn, to conduct experiments and research necessary to employ such an idea. The repository will provide storage for over 300,000 years, when the radioactivity levels of the spent nuclear fuel will reach that of the background radiation and can be designed with the possibility of fuel recovery for further reprocessing. The Swedish government and authorities have approved the plans to establish a final repository for long-lived high-level nuclear waste at Forsmark. In Finland, the process, which was developed in collaboration with SKB in Sweden, has gone further and the final repository is currently under construction on the Olkiluoto island.

While Sweden and Finland have made significant progress in developing a geological repository for spent nuclear fuel, by way of contrast, the Yucca Mountain project in the United States has faced numerous challenges and setbacks. Initiated in the 1980s, the project has been officially shelved since being defunded in 2010. Debate is ongoing over the suitability of the site and the need for a permanent nuclear waste repository in the US. Despite decades of research and study, significant technical concerns, political opposition, and (perhaps most importantly) public trust remain factors against the use of

Yucca Mountain. However, the lack of alternative solutions, coupled with potential political and legal shifts, leaves the door slightly ajar for its potential revival in the future. In the meantime, the US efforts have been focused on generic research and development work that will support future selection of site-specific geologic repositories, among other options [14]. The contrast between the US and the Scandinavian countries vis-à-vis the status of permanent nuclear waste solutions highlights the technical, societal, and economic complexities involved in establishing a permanent disposal solution for high-level radioactive waste.

6. Alternative Uses for Depleted Uranium

Deconversion and recovery processes for depleted UF_6 offer the potential to yield valuable products aside from the recovery of fluorine. One important byproduct of deconversion is the generation of depleted uranium oxide, which can be utilized in various industrial applications [15]. Depleted uranium oxide can be used in the production of radiation shielding materials, such as counterweights in aircraft and ballast in ships. Its high density and effectiveness in absorbing radiation make it valuable in applications where protection against ionizing radiation is necessary. Additionally, depleted uranium metal can be used in the manufacturing of armor-piercing ammunition and as a component in certain types of industrial counterweights and radiation shielding. Furthermore, the recovery of fluorine from the deconversion process can have industrial applications in the production of various chemicals, including inorganic fluorides, fluorocarbons, and fluorine gas. These fluorine-based products are used in a wide range of industries, from pharmaceuticals and agrochemicals to the manufacturing of refrigerants and fluoropolymer materials.

Overall, the deconversion and recovery of depleted UF_6 can contribute to the production of valuable materials with industrial and commercial significance, adding another dimension to the potential benefits of managing and repurposing nuclear byproducts. However, these alternative use cases appear to represent only a minor contribution to the waste management landscape.

7. Potential for Recycling Spent Fuel and Depleted Uranium

In the case of a closed fuel cycle the spent nuclear fuel will be reprocessed and reused for power, heat, electricity production and more. Element separation is employed in spent nuclear fuel reprocessing. The fuel can be separated into a plutonium stream, a uranium stream, a minor actinide stream, and a stream of remaining waste products. Plutonium and uranium combined can be used to power reactors as a reprocessed mixed oxide fuel (MOX). Fuel reprocessing enhances resource utilization, promotes sustainability and reduces the radioactivity and radiotoxicity of the finally disposed spent nuclear fuel. France produces the largest amount of electricity by nuclear power in Europe, over 70%. By employing the spent fuel reprocessing method, it is reported that they require 17% less fresh uranium to operate their plants compared to the one than what would have been needed without reprocessing [16].

Neptunium (Np), americium (Am) and curium (Cm) are called minor actinides (MA). They are produced at small concentrations inside the nuclear fuel as a neutron capture byproduct. A typical light water reactor (LWR) of 1 GWe capacity will generate about 20 tonnes of irradiated fuel annually, out of which 0.1-0.2% is MA. Although relatively small in mass (around 20-25 kg per year) MA contribute disproportionally to the decay heat and radiotoxic hazard of spent nuclear fuel. Therefore, there is a high interest in transmuting them either by fission or neutron capture in order to reduce their impact on the back end of the fuel cycle [17].

Advanced reactor designs are under various stages of design, construction and deployment, worldwide that will provide the potential for efficient utilization of MA in the form of advanced fuel concepts.

8. Next Generation Reactor Technologies

In addition to the potential for recycling spent fuel, there are indeed several next-generation (Generation IV) reactor technologies that could be used for uranium recycling and fuel reprocessing. These advanced reactor technologies offer innovative methods for utilizing spent fuel and depleted uranium, addressing concerns regarding long-term storage and waste management while also providing sustainable options for nuclear energy production.

Generation IV designs comprise reactor concepts that will enhance fuel utilization efficiency, minimize waste production, be economically competitive, incorporate passive safety systems, and promote proliferation resistance. The six Generation IV reactor designs include the: Sodium-cooled Fast Reactor (SFR), Lead-cooled Fast Reactor (LFR), Very High Temperature Reactor (VHTR), Molten Salt Reactor (MSR), Gas-cooled Fast Reactor (GFR) and the Supercritical Water-cooled Reactor (SCWR) [18].

Sodium-cooled Fast Reactors (SFRs) (see Figure 5) use liquid sodium coolant, allowing for high power density and low-pressure operation. Sodium reacts with air and water, which is why the coolant needs to be sealed and isolated from the environment. The SFR can operate on a closed fuel cycle which regenerates fissile fuel and manages minor actinides. Key safety features include long response times, a good margin to coolant boiling, near-atmospheric pressure operation, and an intermediate sodium system for the isolation of the coolant from the electricity producing cycle. The power conversion system can use water, steam, supercritical CO_2, or nitrogen for high efficiency and reliability. The SFR aims to be cost-effective and to extend uranium resources better than thermal reactors. The SFR technology is the most mature Generation IV concept and has been proven through past and current projects in France, Japan, Russia, China, and India.

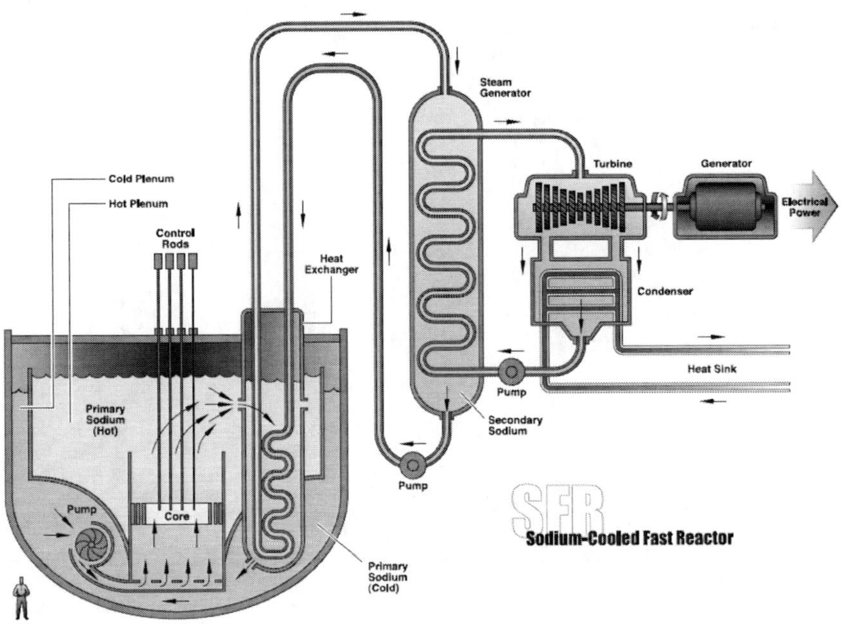

Figure 5. Schematic of a Sodium-cooled Fast Reactor. Image courtesy of the Generation IV International Forum [19].

Lead-cooled Fast Reactors (LFRs) (see Figure 6) operate using fast neutrons and offer the potential for a closed fuel cycle, meaning that they can optimally use the fuel and reduce the waste stream. LFRs can also consume waste from other reactors and could include thorium as fuel. The safety parameters are enhanced by using molten lead as a coolant, which is stable, operates at low pressure, and has a very high boiling point, eliminating the risk of boiling and coolant loss. Lead also helps contain radioactive materials and reduces the risk of chemical reactions with the environment in case of leakage or severe accidents. In terms of sustainability, lead cooling allows for better resource use and passive safety features like natural cooling. The designs can be made simpler and more economical since the reactor can be designed so as to not require additional cooling systems or pressurization. Research questions with this class of designs include mostly issues with structural materials erosion and corrosion as well as reactor core inspection challenges due to the coolant opacity and high temperature shutdown. LFR designs make them resilient to power outages and prevent hydrogen explosions, which were among key issues that contributed to the Fukushima-Daiichi accident.

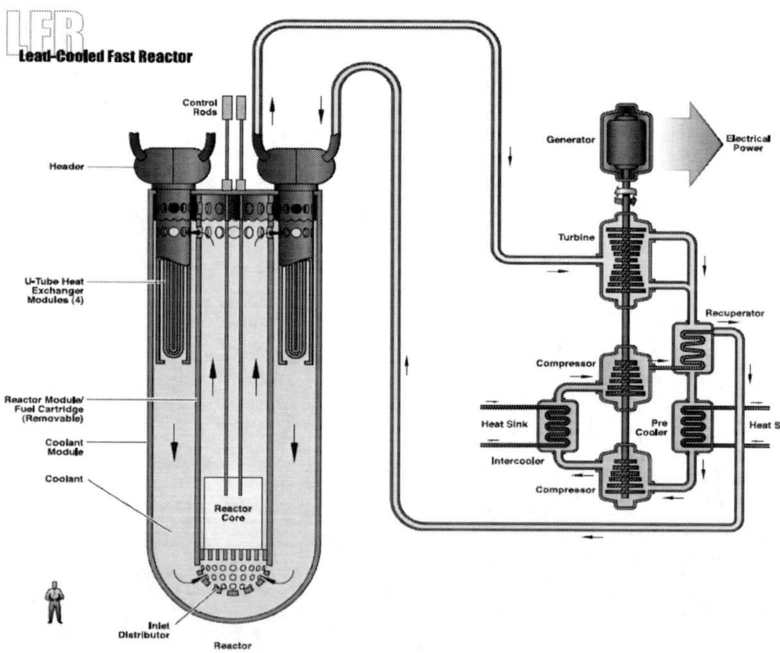

Figure 6. Schematic of a Lead-cooled Fast Reactor. Image courtesy of the Generation IV International Forum [19].

Very High Temperature Reactor (VHTRs) (see Figure 7) are designed for cogenerating electricity and hydrogen. Their high outlet temperatures make them suitable for hydrogen production for industrial applications such as chemical production, pyrolysis of biomass and production of e-fuels. VHTRs utilize TRISO-coated particle fuel [20], a graphite core, and helium coolant. VHTRs initially focused on very high temperatures for hydrogen production but were, after later market assessments, shifted in focus to lower temperatures (700-850°C) due to the more manageable technical risks and broader application potential. This shift has led to greater emphasis on designs like HTR-PM (High-Temperature Reactor-Pebble-bed Module) in China and the NGNP (Next Generation Nuclear Plant) in the US, which balance high efficiency with reduced technical complexity. The design of VHTRs allows for various fuel cycles, including uranium-plutonium, plutonium, mixed oxide, and uranium-thorium, supporting its versatility and long-term viability.

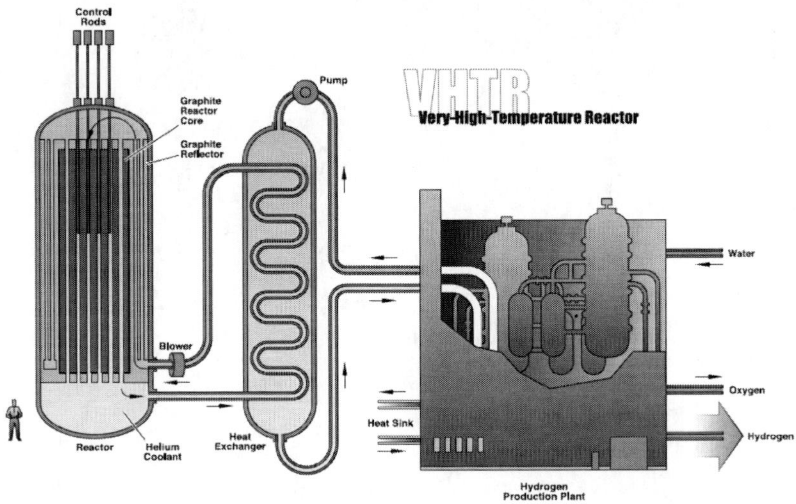

Figure 7. Schematic of a Very-High-Temperature Reactor. Image courtesy of the Generation IV International Forum [19].

Molten Salt Reactors (MSRs) (see Figure 8) use fuel dissolved in molten salts, often in fluoride form. This technology was first explored in the 1960s. MSRs are quite interesting for their fast neutron spectrum and thorium fuel cycle potential. Online fuel reprocessing allows MSRs to breed plutonium or uranium-233 from thorium thus producing more fissile material than what they consume. Key challenges include developing safety approaches, corrosion control, significant radiation damage to structural components, improving online salt processing, and providing a strong proliferation resistance. MSRs provide lower fissile inventories, no need for solid fuel fabrication, and a homogeneous isotopic composition of fuel, making them effective for actinide burning and extending fuel resources. Interest in liquid salt as a coolant has increased for both nuclear and non-nuclear applications, including nuclear hydrogen production and solar electricity generation. This highlights the potential versatility and efficiency of MSRs.

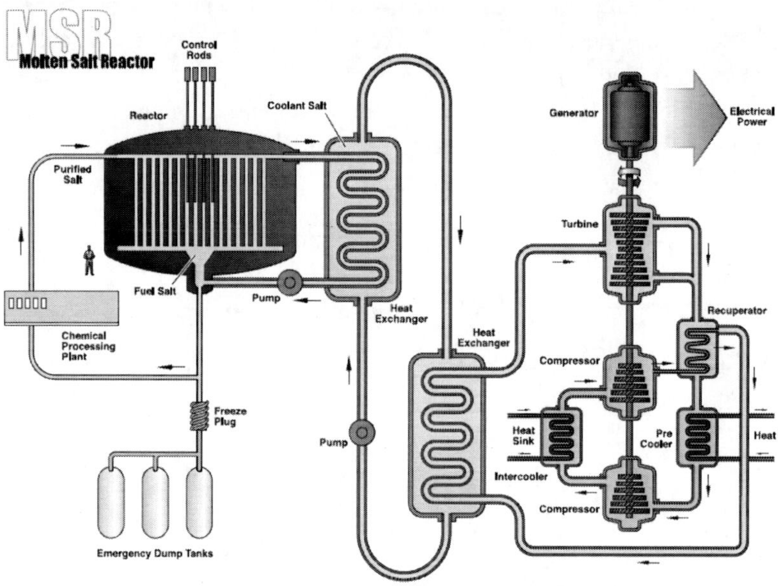

Figure 8. Schematic of a Molten Salt Reactor. Image courtesy of the Generation IV International Forum [19].

Gas-Cooled Fast Reactors (GFRs) (see Figure 9) are high-temperature, helium-cooled reactors designed for long-term sustainability and efficient waste management. They are conceptually designed to use a closed fuel cycle to recycle fuel and minimize waste while operating at high temperatures for improved thermal efficiency. GFRs combine features from existing reactor technologies: they share fuel recycling processes with SFRs and similar structural and material technologies from Very High-Temperature Reactors (VHTRs). However, they require additional research on core design and safety. GFRs have a similar thermal setup to combined cycle systems in natural gas plants and also includes a steam generator to further utilize waste heat.

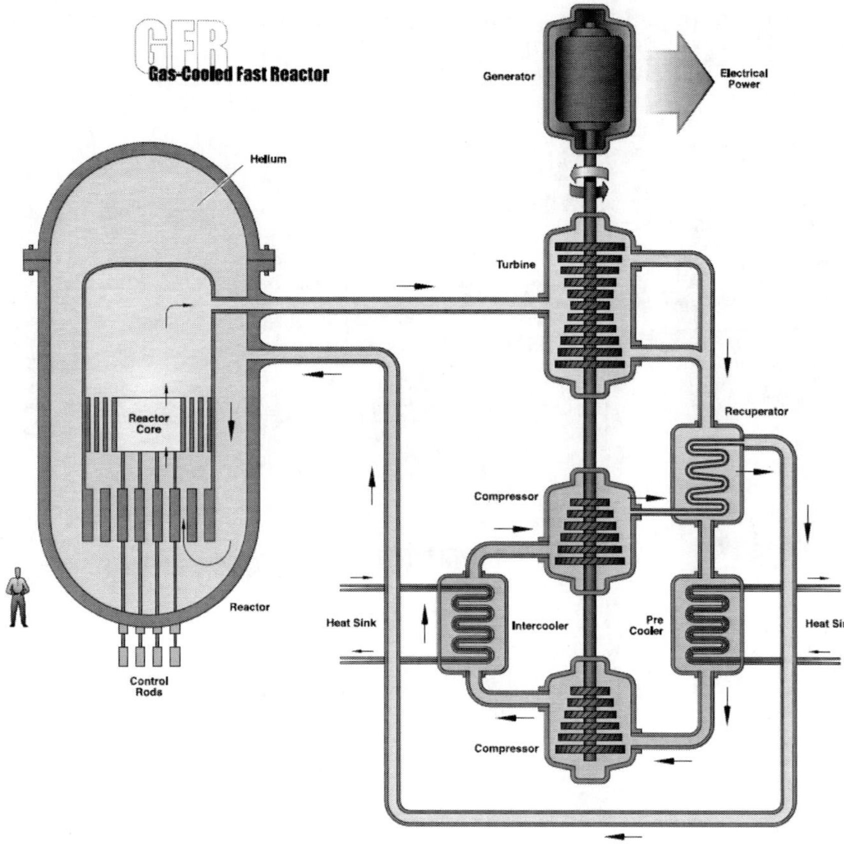

Figure 9. Schematic of a Gas-cooled Fast Reactor. Image courtesy of the Generation IV International Forum [19].

Supercritical Water Reactors (SCWRs) (see Figure 10) are high-temperature, high-pressure water-cooled reactors that operate above water's critical point. They can use either a thermal or fast-neutron spectrum, depending on the core design, and may use light or heavy water as a moderator. SCWRs offer several advantages over current reactors such as enhanced efficiency, up to 44% compared to 31-36% for current reactors, simplified design, no need for reactor coolant pumps or traditional steam generators, reducing capital and operational costs, as well as reduced size, with smaller containment offering lower costs. However, SCWRs face challenges including validating heat transfer models, qualifying advanced materials, and demonstrating effective passive safety systems. SCWRs are currently being developed in Japan, Europe, Canada, Russia and China. Research continues on materials, especially for high temperatures, and water chemistry to address corrosion and stress corrosion cracking.

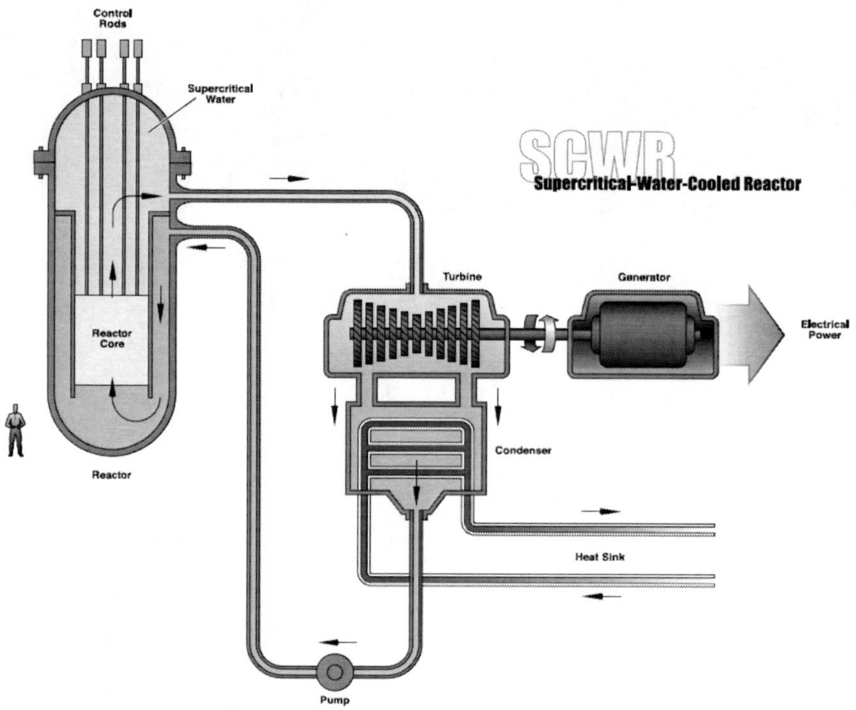

Figure 10. Schematic of a Supercritical-Water-cooled Reactor. Image courtesy of the Generation IV International Forum [19].

9. Outlook on Gen IV Technologies

Nuclear power plants, which produce low-carbon electricity at stable and competitive costs, constitute an element of the solution to global climate change and a means of delivering power to emerging and developed countries. The global nuclear power plant fleet is aging, with a wave of reactors scheduled for decommissioning in the coming decades. Over 85% of nuclear power plants currently in operation would be classified as “Gen II” reactors, most of which are close to (or have exceeded) their initially-intended 30-40 year operational lifespan [21]. Many operators are pursuing life extension programs to maximize the operating life of existing plants out to 60, or perhaps even 80 years [22]. These extensions are possible because of the inherent safety margins built into the initial designs, coupled with refurbishments and upgrades to further enhance operational safety. However, this trend toward operational life extension must yield to new construction should nuclear power continue to play a role in the energy landscape beyond the next half-century. Further development of nuclear technology is needed to meet future energy demands.

The 2011 Fukushima Daiichi nuclear power plant accident underscored the importance of designing nuclear systems with the highest safety standards. It also rekindled public fears about nuclear safety, bringing back memories of the 1979 Three Mile Island and 1986 Chernobyl accidents. Despite significant safety improvements and stricter regulations in the nuclear industry, a negative public perception and fear of nuclear accidents remain in some regions. This remains a significant obstacle to the full development and implementation of Gen IV technologies.

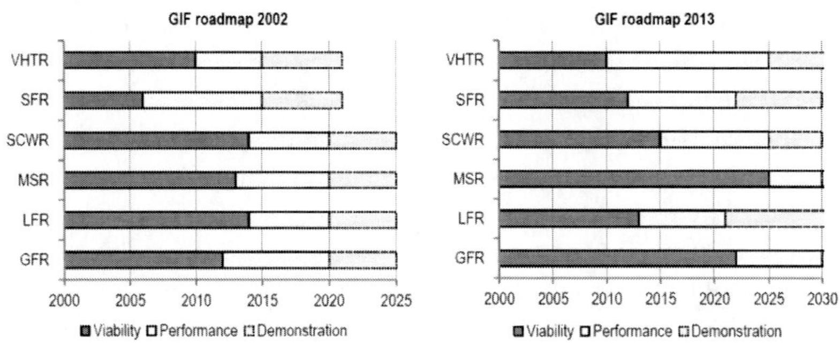

Figure 11. Projected Gen IV reactor development timelines. Image courtesy of the Generation IV International Forum [23].

Lessons learned from Fukushima, particularly the need for reliable long-term residual heat removal and the prevention of off-site releases during severe accidents, are being applied to both the current operating fleet and future nuclear systems, including Generation IV designs. These advanced reactors aim to improve upon existing designs by incorporating features such as non-water coolants, higher operational temperatures and power densities, and in some cases, integrated fuel cycle facilities.

The Generation IV International Forum (GIF), a consortium of nuclear-capable countries, provides guidance and shares information to demonstrate the capabilities of these advanced systems. The GIF has developed "roadmaps" outlining development and deployment of six Gen IV reactor designs. While the initial 2002 timeline proved optimistic, the 2013 roadmap offered a more realistic outlook, acknowledging delays and adjusting timelines accordingly (see Figure 11) [23]. In the current landscape, SFR and VHTR technologies are the most mature. However, beyond addressing the technical and research challenges needed for each of the Gen IV technologies, securing funding, political will, and public acceptance will be crucial for the widespread deployment of Gen IV reactors to become a reality.

Acknowledgments

WAA is grateful to be joined by EC and PO in the collaboration of this chapter. EC is also known as @YourFriendlyNuclearPhysicist on YouTube, where the interested reader may find further information on many topics of nuclear physics and the nuclear industry.

Preparation of the manuscript was supported by a Summer Faculty Stimulus Grant from the College of Arts and Sciences at The University of Memphis. The author additionally thanks April, Kennedy, Liam, and Logan for their support of this work. Any opinions, findings, and conclusions or recommendations expressed in this book chapter are those of the author(s) and do not necessarily reflect the views of the funding/supporting agencies.

Figures in this chapter have undergone graphical alterations/edits to conform to publication standards and are intended for use in this chapter as *illustrations*. The original referenced figures should be used for rigorous scientific examination/comparison.

This work would not have been possible without the intellectual contributions of Dr. Megan E. Bennett.

References

[1] *US Nuclear Regulatory Commission,* Conventional Uranium Mills, (Dec 2020) available at https://www.nrc.gov/materials/uranium-recovery/extraction-methods/conventional-mills.html, 2020.

[2] *US Nuclear Regulatory Commission,* Uranium Recovery (Extraction) Methods, (Dec 2020) available at https://www.nrc.gov/materials/uranium-recovery/extraction-methods.html, 2020.
[3] OECD Nuclear Energy Agency/IAEA, *Uranium 2022: Resources, Production and Demand*, OECD Publishing, Paris, 2023.
[4] *US DOE Energy Information Administration,* 2022 Domestic Uranium Production Report, (May 2023) PDF available at https://www.eia.gov/uranium/production/annual/archive/dupr2022.pdf, 2023.
[5] Arcuri, L., Pietrelli, L., & Rizzello, C., Experience with a uranyl nitrate/uranium dioxide conversion pilot plant *IAEA Report* (ENEA-RT-COMB--84-9). Italy, 1984.
[6] Morel, B., Selmi, A., Moch, L., Hiltbrunner, J.-M., Achour, M., Benzouaa, R., Bock, A., Jouffret, L., Bonnet, P., Hamwi, A., Dubois, M. Surface reactivity of uranium hexafluoride (UF_6), *Comptes Rendus Chimie* 2018, *21*, 782-790.
[7] *US Nuclear Regulatory Commission*, Backgrounder on Uranium Enrichment, (Dec. 2020) PDF available at https://www.nrc.gov/docs/ML0313/ML031330159.pdf, 2020.
[8] Hore-Lacy, I., "Uranium Mining, Processing, and Enrichment," in *Encyclopedia of Energy*, 2004, 317-318.
[9] *World Nuclear Association,* "Uranium Enrichment" (Oct. 2022) available at https://world-nuclear.org/information-library/nuclear-fuel-cycle/conversion-enrichment-and-fabrication/uranium-enrichment, 2022.
[10] Hung, N. T., Thuan L. B., Thuy N. T., Than H. S., Van Phuc D., Lee J. Y., Jyothi R. K. Uranyl ammonium carbonate precipitation and conversion into triuranium octaoxide. Heliyon. 2024 Feb 10;10(4):e25930. doi: 10.1016/j.heliyon.2024.e25930. PMID: 38384576; PMCID: PMC10878953.
[11] H. Stehle, H. Assmann, F. Wunderlich, Uranium dioxide properties for LWR fuel rods, *Nuclear Engineering and Design*, 1975, *33*, 230-260.
[12] IAEA, *Guidebook on Quality Control of Water Reactor Fuel*, Technical Reports Series No. 221, IAEA, Vienna, 1983.
[13] Hedin, A. 2006. *Long-term safety for KBS-3 repositories at Forsmark and Laxemar - a first evaluation. Main Report of the SR-Can project.* Sweden, http://www.skb.se/upload/publications/pdf/TR-06-09webb.pdf, 2006.
[14] Faybishenko, T., B., Birkholzer, J., Sassani, D., and Swift, P., *International Approaches for Nuclear Waste Disposal in Geological Formations: Geological Challenges in Radioactive Waste Isolation - Fifth Worldwide Review*. United States. https://doi.org/10.2172/1353043, 2017.
[15] OECD Nuclear Energy Agency/IAEA, *Management of Depleted Uranium*, OECD Publishing, Paris, 2001.
[16] IAEA, *Management of Spent Fuel from Nuclear Power Reactors*, IAEA Bulletin, vol 60-2, (June 2019), https://www.iaea.org/publications/magazines/bulletin/60-2, 2019.
[17] OECD, *Minor Actinide Burning in Thermal Reactors: A Report by the Working Party on Scientific Issues of Reactor Systems*, Nuclear Science, OECD Publishing, Paris, https://doi.org/10.1787/9789264208537-en, 2014.

[18] OECD Nuclear Energy Agency/Generation IV International Forum, *GIF 2023 Annual Report*, (July 2024); PDF available at https://www.gen-4.org/gif/jcms/c_224113/gif-2023-annual-report, 2024.

[19] U.S. DOE Nuclear Energy Research Advisory Committee/Generation IV International Forum, *A Technology Roadmap for Generation IV Nuclear Energy Systems*, Report GIF002-00, (Dec. 2002); PDF available at https://www.gen-4.org/gif/jcms/c_9352/technology-roadmap, 2002.

[20] Kenzhina, I., Blynskiy, P., Kozlovskiy, A., Begentayev, M., Askerbekov, S., Zaurbekova, Z., Tolenova, A. Study of Morphological, Structural, and Strength Properties of Model Prototypes of New Generation TRISO Fuels, *Materials* 2022, *15*, 4741.

[21] *IAEA,* Nuclear Power Status 2021, PDF available at https://pris.iaea.org/pris/PRIS_poster_2021.pdf, 2021.

[22] Fernández-Arias, P., Vergara, D., Orosa, J. A. A Global Review of PWR Nuclear Power Plants. *Appl. Sci.* 2020, *10*, 4434. https://doi.org/10.3390/app10134434.

[23] OECD Nuclear Energy Agency/Generation IV International Forum, *Technology Roadmap Update for Generation IV Nuclear Energy Systems*, (Jan. 2014); PDF available at https://www.gen-4.org/gif/jcms/c_60729/technology-roadmap-update-2013, 2014.

Chapter 2

Uranium Dioxide: Powder Synthesis, Pressing, and Sintering

Balakrishna Palanki*
(Retired) Nuclear Fuel Complex, Hyderabad, India

Abstract

The characteristics of UO_2 powder made through the aqueous chemical route are greatly dependent on the precursor precipitation conditions such as reactant concentrations, pH, temperature, and agitation. If these conditions are not chosen correctly, the powder will not possess the required ceramic activity and lead to low sintered density. At the other extreme, the powder will be hyperreactive and become agglomerated, or the resulting UO_2 will become oxidized. The agglomerates gain strength during the drying and calcination processes, where weak bonds are replaced by solid bridges. It has been industrial practice to modify the powder to suit pelletizing operations, instead of the other way around. Since it is difficult to make powder that is entirely free from agglomerates, the efforts are usually directed at making soft powders, where the agglomerates tend to separate into individual particles. In addition, UO_2 powder from a fluoride source is milled to reduce the size of the hard agglomerates to acceptable levels.

* Corresponding Author's Email: palankibalakrishna@yahoo.com

In: The Chemistry of Uranium
Editor: William A. Alexander
ISBN: 979-8-89530-802-8

1. Introduction

Concerns about global climate change have revived the importance of nuclear power. Uranium dioxide is the most used compound of uranium, as nuclear power reactor fuel. This chapter contains a compilation of information from published literature. Its goal is to make available in one place information useful to the student or the knowledge worker. Mainly, powder synthesis, pelletizing, and sintering of uranium dioxide have been covered. Summary topic boxes are provided at the beginning of each section. Beyond directly cited literature, an extensive bibliography has been added for the advanced learner.

2. Powder Characterization

The powder characteristics give an idea of the behavior of the powder during processing and the likely product characteristics. BET specific surface area, bulk density, average particle size, particle size distribution etc. are commonly used to characterize the powder. Correlations between powder characteristics and product properties may be generated from small-scale, well-controlled experiments. Thermogravimetry may be used to study the decomposition and calcination of ammonium diuranate. The laser particle size analyzer gives an idea of the particle size distribution. The pore sizes in powder and the open porosity in green pellet form may be investigated using mercury intrusion porosimetry. The sintering behavior may be investigated using a dilatometer, which gives shrinkage as a function of temperature. However, it is risky to accept a powder based on powder characteristics alone. Instead, ceramic powders are accepted based on a sinterability test or a preproduction trial on a representative sample of the powder blend.

2.1. Sieve Analysis

The simplest way to characterize a powder is by sieve analysis. Wet sieving results are more reproducible as the particles get separated from each other in the fluid medium; dry sieving may cause agglomeration of fluffy powders. However, disposal of the radioactively contaminated fluid will have to be taken care of in the case of wet sieving of UO_2 powders.

2.2. BET Specific Surface Area

The driving force for sintering is the reduction in surface energy. The starting powder possesses a high surface energy by virtue of a very small particle size. The BET specific surface area is representative of the surface energy, with a high BET value corresponding to a small particle size. Particle-to-particle contacts are formed in pressing, and the particle boundaries transform to grain boundaries while sintering. BET values of 2 to 4 m^2/g depending on the process route are common for UO_2 powders. The sintered density obtained is expected to increase as the BET value increases.

2.3. Average Particle Size

Average particle size (APS) is determined using a Particle Size Analyzer, such as the Fisher Sub Sieve Sizer (FSSS). The FSSS uses pressure drop across a powder volume for determining the particle size. When the powder is made up of fine particles, the drop is larger than when it is made up of coarse particles. The particle size indicated by the method is usually much larger than that calculated from BET specific surface area. This indicates that some particle clusters exist and are not permeable to gas flow. APS of < 2 μm by FSSS is common for UO_2 powders. The sintered density is expected to increase as the APS is decreased.

2.4. Particle Size Distribution

Figure 1 is an example of a particle size distribution of UO_2 powder obtained by the laser diffraction method. This sample has shown a bimodal distribution.

The nature of the open pores in a powder or green body can be inferred from mercury porosimetry. Mercury is used because it is nonwetting to both glass and the materials being investigated (high contact angle). A powder or green pellet sample is placed in the instrument and pressure evacuated. Then mercury is introduced into the sample, and the chamber is pressurized. Initially, the largest pores in the sample fill with mercury. As the pressure is increased, the mercury will be able to penetrate smaller pores. The volume of the mercury as a function of applied pressure gives an idea of the pore size distribution. Figure 1 gives the pore size distributions of UO_2 ex-ADU powder, and of green pellets with densities of 4.94 g/cc and 5.88 g/cc.

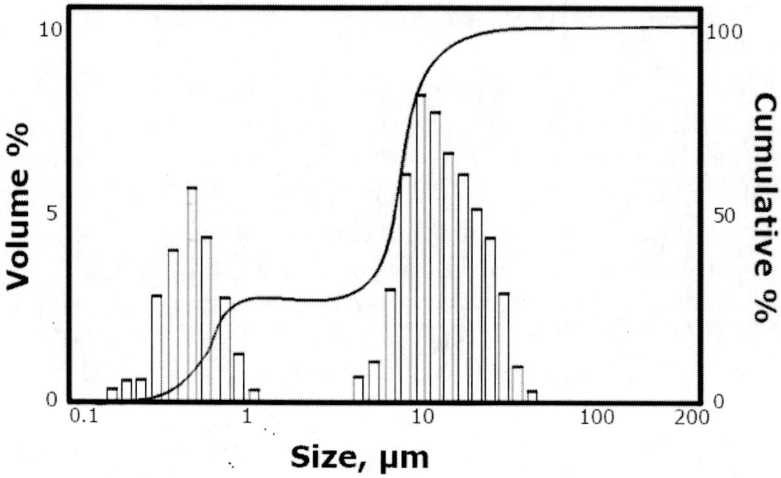

Figure 1. An example particle size distribution by Laser Diffraction. Particle size in µm is indicated on the logarithmic *x*-axis. The volume percent of each size is indicated on the *y*-axis in the form of vertical bars. The solid curve indicates cumulative volume percent of particles versus particle size.

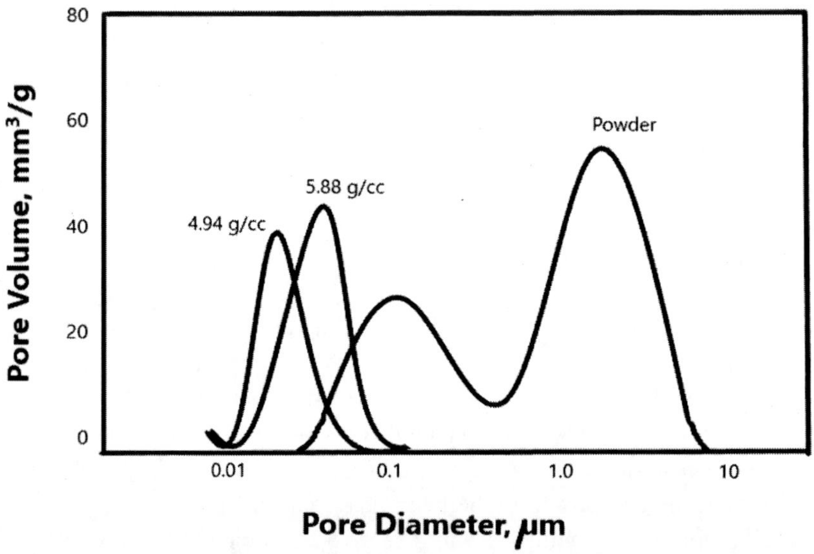

Figure 2. Pore Size Distribution of UO_2 ex ADU, green pellet of 4.94 g/cc and of 5.88 g/cc [1].

2.5. Dilatometry

A dilatometer is a precision instrument for the measurement of dimensional changes in materials as a function of temperature. For example, one can measure thermal expansion of materials and sintering shrinkage of a powder compact as its temperature is raised. Dilatometry can be used to test a wide range of materials including traditional and advanced ceramics, glasses, metals, and polymers. Figure 3 shows the sintering linear shrinkage of UO_2 compacts derived from different powder routes as observed in a dilatometer.

2.6. Scanning Electron Microscopy

Scanning electron microscopy is used to understand the physical nature of the powder and pellet sections. Figure 4 (a-c) displays scanning electron microphotographs of UO_2 powder from ADU (ammonium diuranate), AUC (ammonium uranyl carbonate) and IDR (Integrated Dry Route) routes. The crystalline nature of UO_2 ex-AUC can be appreciated in Figure 4b.

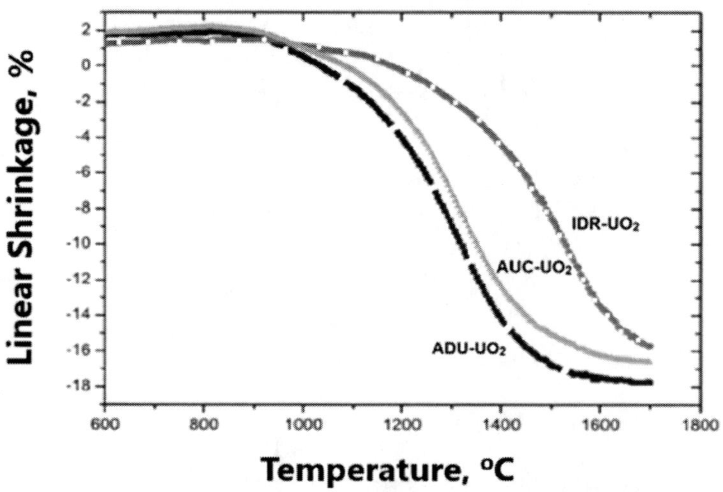

Source: With kind permission from Korean Nuclear Society.

Figure 3. Sintering Shrinkage of UO_2 Compacts Derived from Different Powder Routes as observed in a dilatometer [2]. Of the three powder compacts, the one from ADU has sintered fastest, while the one from IDR required higher temperatures for the same linear shrinkage. The compact from AUC has exhibited intermediate behavior.

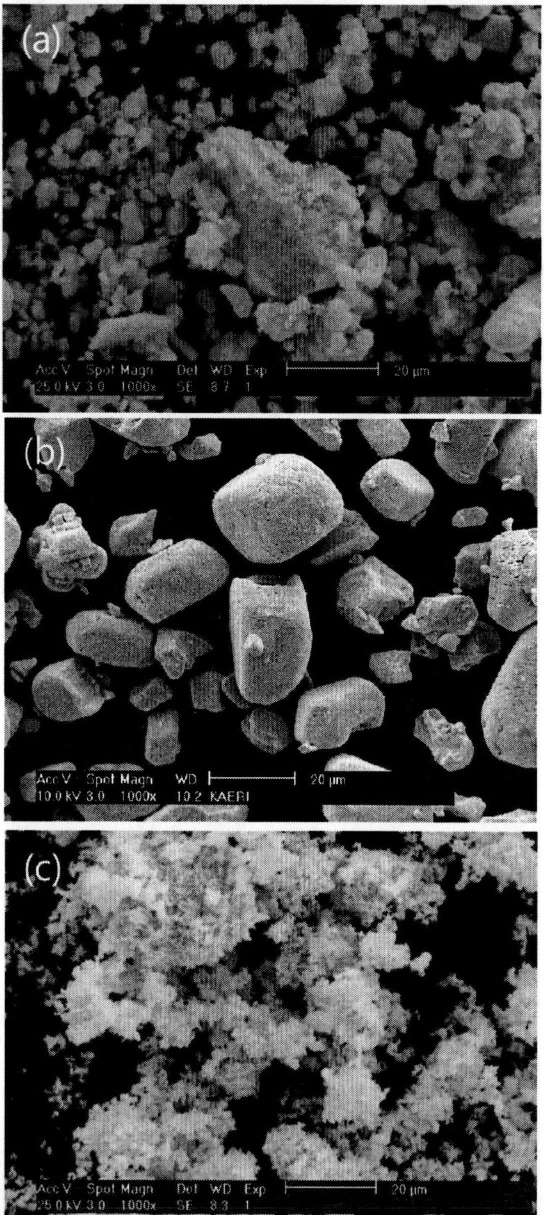

Source: With kind permission from Korean Nuclear Society.

Figure 4. UO_2 powder from (a) ADU, (b) AUC, and (c) IDR [2].

3. Powder Synthesis

Fine powders suitable for pressing and sintering are made through several industrial processes, one of which is the aqueous chemical route involving precipitation. Precipitation is a rapid process of nucleation and growth when compared to crystallization. Depending on the conditions of precipitation, the precipitate formed may be crystalline, curdy, or gelatinous. In the aqueous chemical route for the preparation of sinterable powders, the aim is to produce filterable (or decantable, or centrifugable) crystalline precipitates. Usually, there is a trade-off between the filterability of the precursor precipitate and sinterability of the calcined oxide powder. Two-stage precipitation with intense stirring enables formation of a fine, filterable precipitate while ensuring maximum recovery of the uranium from solution. The precipitate morphology is retained in the final powder. The thermal treatment parameters (rate of heating, soaking temperature, and time) are selected to create porosity in the precipitate precursor without the loss of precipitate morphology and without free sintering or formation of strong inter-particle bridges. The final powder from a precursor porous morphology is soft, agglomerate-free, and requires only moderate pressing pressures.

3.1. Powder Flowability, Formability and Sinterability

Beach sand flows readily but it can neither be compacted nor sintered. On the other hand, clay particles do not flow readily, but they can be shaped and sintered (See Figure 5), as in pottery, a technology that existed thousands of years ago. Differences in particle behavior arise from differences in particle sizes and particle size distributions, particle shape, density, etc. For sintering, the starting powder must be very fine. Fine UO_2 particles possess a high specific surface area of the order of 2-4 m^2/g.

Figure 5. Two powders - beach sand and potter's clay, with different forming and firing characteristics.

3.2. Crystallization and Precipitation

In crystallization and precipitation, there are three basic steps: nucleation, the formation of initial crystal nuclei; crystal growth, the subsequent growth of these nuclei into microcrystals; and aggregation, the joining of the micro crystals to form larger entities. The aggregates can further combine into larger secondary and tertiary agglomerations. The key differences between the crystallization and precipitation processes are discussed below [3].

In crystallization, the process of crystal growth (rather than aggregation) dominates. There are fewer nuclei that are made to grow into large crystals over a period. Crystallization is reversible. The crystal re-dissolves, for example, on increasing the temperature. In contrast, precipitation is an irreversible process and requires acidification to re-dissolve.

In precipitation, growth by aggregation (rather than crystal growth) dominates. Due to instant super saturation, there are several nuclei formed that quickly aggregate. The tendency to aggregate may be countered to some extent by intense stirring. Even though crystal growth is small, it may be possible to increase its extent. The reason for aiming for a little growth of the micro crystals is to decrease the tendency to aggregate. Precipitation using dilute solutions under 'low' pH and 'high' temperature favor growth and lead to denser micro crystals with a lower prospect of aggregating.

3.3. Types of Precipitates

The morphology or physical appearance of precipitates is determined by nucleation and growth rates, colloidal stability, recrystallization, and aging processes, and various habit modifications brought about either by ions or molecules present in the solution or the solvent itself. Thus, by controlling these parameters, precipitates can be produced with either the properties of well-formed, relatively large crystallites or, at the other extreme, amorphous, or even gel-like products. In other words, any precipitate can be obtained in any desired form if the right conditions are chosen (rate of reactant addition, stirring efficiency, temperature, pH etc.). With a slow, controlled growth of sparingly soluble salts, it is often feasible to produce monodisperse precipitates, that is, precipitates containing particles of near equal size. The elementary ADU precipitate particles are usually very thin platelets. These platelets are normally attracted to each other, plane to plane, by weak inter-molecular forces [4] and are subsequently bonded by chemical forces, leading to the formation of particles with a 'woven basket' structure [5] and stable aggregates of the particles. The finer the precipitate particle, the greater the tendency is to agglomerate. The agglomerates strengthen and secondary agglomerates may be formed during precipitation and subsequent thermal treatments.

3.4. Precipitation of Ammonium Diuranate (ADU)

Sinterable high purity fine powders for nuclear fuel are usually made through aqueous chemical processing. ADU may be precipitated either from uranyl fluoride or uranyl nitrate.

3.4.1. Fluoride Route

Precipitation from a uranyl fluoride solution is used when the starting material is low-enriched UF_6 in the manufacture of light-water reactor (LWR) fuels.

Three chemical equations are of interest.

Hydrolysis: $UF_6 + 2\ H_2O = UO_2F_2 + 4\ HF$

Neutralization of 'free' acid: $HF + NH_4OH = NH_4F + 3\ H_2O$

Precipitation: $2\ UO_2F_2 + 6\ NH_4OH = (NH_4)_2U_2O_7 \downarrow + 4\ NH_4F + 3\ H_2O$

While the formula $(NH_4)_2U_2O_7$ suggests an ideal 1:1 ratio of NH_3:U, the actual ratio in the precipitates is often less than 1 and depends on the precipitation conditions. The precipitate is therefore a *mixture* of non-stoichiometric ammonium polyuranates, collectively referred to as ammonium diuranate (ADU).

3.4.2. Nitrate Route

ADU is also precipitated from uranyl nitrate. Nitrate solutions are used in the manufacture of CANDU (CANada Deuterium Uranium, i.e., heavy-water cooled reactor) fuel and in the processing of scrap low-enriched uranium in LWR fuel.

Neutralization of 'free' acid: $HNO_3 + NH_4OH = NH_4NO_3 + H_2O$

Precipitation: 2 $UO_2(NO_3)_3$ + 6 NH_4OH = $(NH_4)_2U_2O_7$ ↓ + 4 NH_4NO_3 + 3 H_2O

To precipitate all the uranium from the solution, it is necessary to add a greater amount of ammonium hydroxide (or ammonia gas) than stoichiometrically indicated. The excess NH_3/U excess mole ratio required is significantly higher for precipitation from uranyl fluoride than from uranyl nitrate solutions. This difference arises from the greater stability of uranium fluoride complexes.

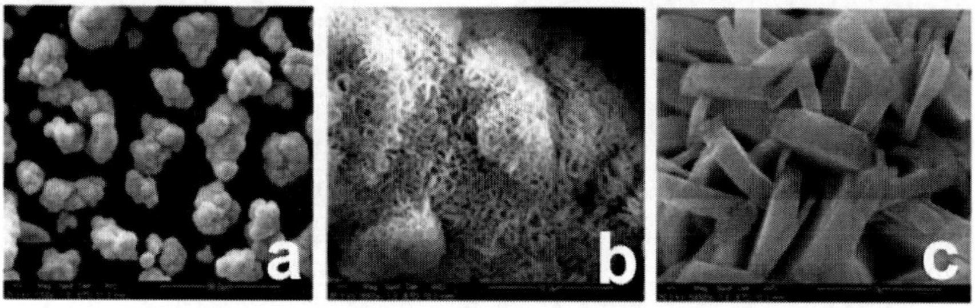

Figure 6. SEMs of ammonium diuranate (a) 500X (b) 5000X (c) 50000X [5]. The woven basket structure is seen in (b).

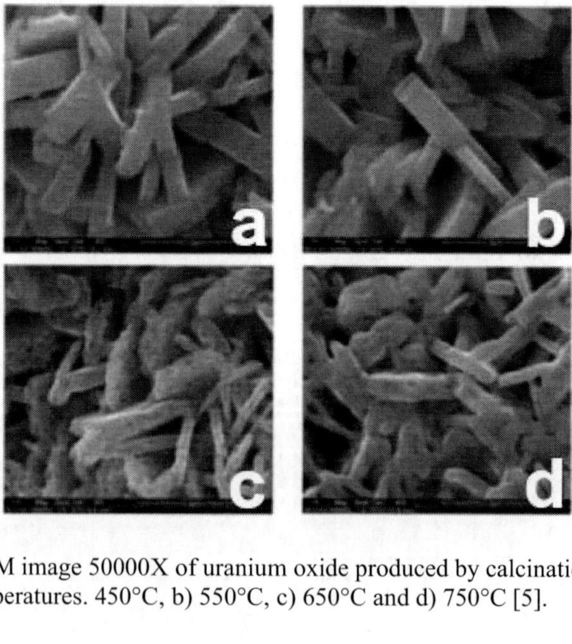

Figure 7. SEM image 50000X of uranium oxide produced by calcination of ADU at different temperatures. 450°C, b) 550°C, c) 650°C and d) 750°C [5].

Figure 6 shows Scanning Electron Microphotographs of ammonium diuranate at three different magnifications. The microcrystalline nature is visible at 50000 magnification.

ADU is said to be microcrystalline, as the crystal features are discernible at high magnifications, as seen in Figure 6c. Each aggregate shown in Figure 6a is made up of smaller clusters, which are, in turn, made up of micro- or nanocrystals.

Figure 7 shows the SEMs of uranium oxide powder obtained by calcining ADU at different temperatures.

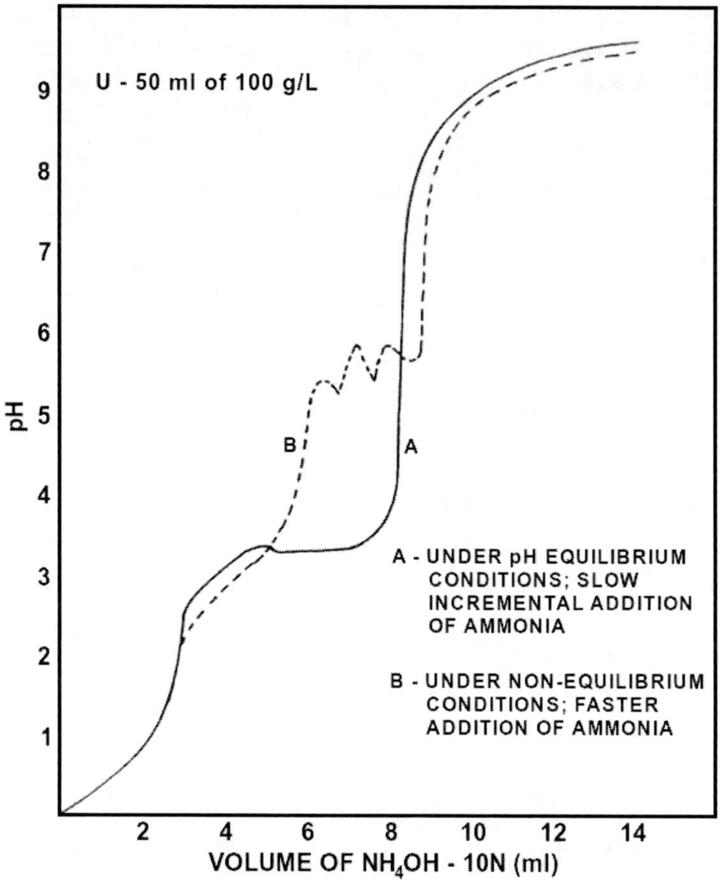

Figure 8. Potentiometric titration curve for uranyl nitrate versus ammonium hydroxide [6].

When uranyl nitrate is titrated with ammonium hydroxide, the pH initially rises as free acid is neutralized. As ADU precipitates, the pH remains somewhat steady. Once precipitation is complete, the pH rises rapidly again but then levels off despite continued addition of ammonium hydroxide, due to the buffering action of ammonium nitrate. Figure 8 shows a typical titration curve for uranyl nitrate with ammonium hydroxide, highlighting this buffering effect. Figure 9 shows the two-stage precipitation set up for uranyl nitrate. The ideal process employs this two-stage precipitation with ammonium hydroxide at low pH and then high pH.

The first stage utilizes a dilute NH_4OH solution (e.g., 2N) and aims to form a filterable (or decantable) precipitate. The second stage uses a more concentrated NH_4OH solution and aims at the recovery of uranium. Both precipitation vessels use recirculation along with intense agitation with mechanical stirrers to break up agglomerates of ADU.

The effect of ammonia flow rate and precipitation temperature on the precipitate characteristics (e.g., particle size, density) is shown as an example in Table 1 [8]. It was found that a low flow rate of NH_4OH and low temperature resulted in minimum agglomeration and the highest bulk density in the ADU precipitate. The higher flow rate resulted in agglomeration.

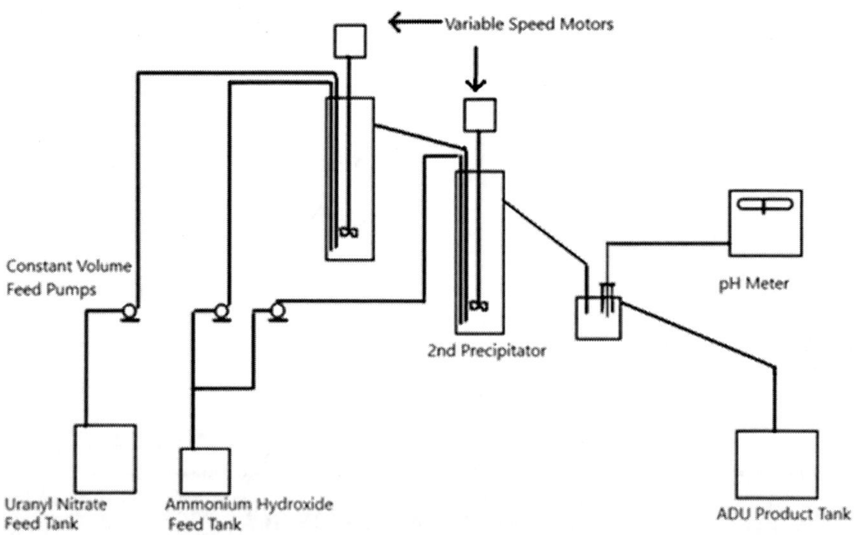

Figure 9. Continuous precipitation of ADU from uranyl nitrate with two precipitation vessels [7].

Table 1. Characteristics of dried ADU versus ammonia flow rate and temperature ([U] = 120 g/l; free acidity = 0.25 N; [NH_4OH] = 7.4 N)

	Characteristic	NH_3 flow rate 26 ml/h		NH_3 flow rate 87 ml/h	
		30°C	60°C	30°C	60°C
1	SSA, m^2/g	0.68	9.29	1.48	8.89
2	Bulk Density, g/cc	1.936	0.939	1.628	0.904
3	Tap Density, g/cc	2.304	1.269	2.153	1.256
4	Plus 325#	1.6	24	10	84
5	Agglomeration number*	10	24,629	55	10,196

*Randall German defined agglomeration number as $N_A = (D\rho S)^3/2620$, where D is the agglomerate median size (μm), ρ is the pyknometric density (g/cc) and S is the BET specific surface area (m^2/g). In this Table, laser particle size distribution was used to determine the median diameter.

3.5. Decomposition of ADU and Calcination

The characteristics of precursor material ADU are inherited into the final product UO_2 powder. Industrial productions are mostly based on two routes, either via uranyl nitrate with aqueous or gaseous ammonia, or via the reaction of uranyl fluoride with ammonium hydroxide. Since the properties of the final product depend on the starting material, it is important to understand the decomposition steps and the intermediate products of ammonium diuranate.

The starting material ADU or $2UO_3.NH_3.3H_2O$ undergoes a process involving several endothermic and exothermic reactions. In situ HT-XRD shows that amorphous UO_3 is obtained after achieving complete dehydration at 300°C, and denitration at about 450°C. After cooling from heat treatment at 600°C, a crystalline UO_3 phase appears, as displayed by ex situ XRD. The self-reduction of UO_3 into orthorhombic U_3O_8 takes place at about 600°C, but a long heat treatment or higher temperature is required to stabilize the structure of U_3O_8 at room temperature. U_3O_8 remains stable in air up to 850°C. Above this temperature, oxygen losses lead to the formation of U_3O_{8-x}, as demonstrated by subtle changes in the diffraction pattern and by a mass loss recorded by TGA [9].

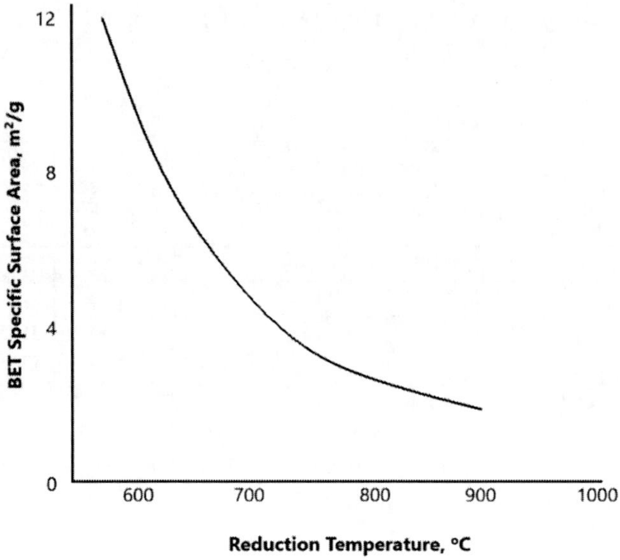

Figure 10. Specific surface area of UO_2 powders reduced at various temperatures in hydrogen [10].

Figure 10 shows the effect of thermal treatment temperature on the specific surface area of UO_2 powder. The specific surface area is chosen carefully, since low specific surface area powders yield low sintered densities, while high specific surface area powders may yield pyrophoric powders. High surface area powders agglomerate and depress sintered densities. Even bloating may occur in the sintered pellet.

When subjected to increasing thermal treatment, the nature of the calcined powders changes from a poor to a well-developed crystalline modification as revealed from X-ray diffraction studies [11].

3.6. Soft and Hard Powders

Soft ceramic powders are those which require low or moderate forming pressures, while hard powders necessitate high-forming pressures. High-forming pressures are undesirable because they can induce stress-relief defects in formed bodies and accelerate the wear of tooling and machinery. Furthermore, high compaction pressure reduces the number of compacts produced per pressing cycle, as the available force must be distributed among all the punches. The powder hardness can stem from the presence of strong, coarse particles or strong inter-particle bonds within agglomerates or particle clusters. Soft powders pack readily in the pressing die at moderate forming pressures because the powder particles readily separate, deform, or fragment, effectively filling the voids between them. Soft agglomerates deform easily during pressing and become more uniform, resulting in green compacts with near monomodal pore spectra.

The original solid particles obtained via precipitation (of size circa 30-56 μm) retain their shape and size but become porous upon controlled thermal decomposition. The porosity develops as the gaseous decomposition products escape from the particle. The greater the porosity, the softer the powder. Many smaller crystals (i.e., of a size < 1 μm) of the new phase appear within the framework of the original precursor particle. If the precipitate is ammonium diuranate (ADU), water and ammonia (or, nitrogen and hydrogen) leave the crystal during drying and calcination, and the atoms rearrange to form nuclei of U_3O_8/UO_2 which then grow.

In a soft powder, the smaller crystals of the new phase are not strongly held together. They can separate and move independent of each other during compaction. The same can be said about particles within an agglomerate. But when the powder is subjected to excessive thermal treatment, bonds develop between the particles. A powder is hard if the constituent grains are not able to separate and rearrange in compaction. A hard powder leads to inefficient packing, which leaves voids that do not close in the sintering process. A soft powder leads to higher packing efficiency in compaction, and homogeneous green and sintered microstructures. Powder softness is lost during rapid or excessive thermal treatment (that is, drying, calcination, and reduction) where the original porosity developed during thermal decomposition is not preserved. Very fine particles tend to join to form agglomerates that strengthen during thermal treatment to form grit. The agglomerates make powder packing inefficient and leave discontinuities in the green body. The discontinuities in the green body may become more pronounced after sintering. Process conditions may be optimized for agglomerate-free fine UO_2 powders [12].

In the case where UO_2 powder is made from UF_6, the residual fluoride causes formation of hard agglomerates with strong bridges between particles. Powders made from the nitrate route are relatively free from agglomerates.

Figure 11 details the flow sheet for production of UO_2 powder starting with magnesium diuranate (MDU) and following the nitrate route.

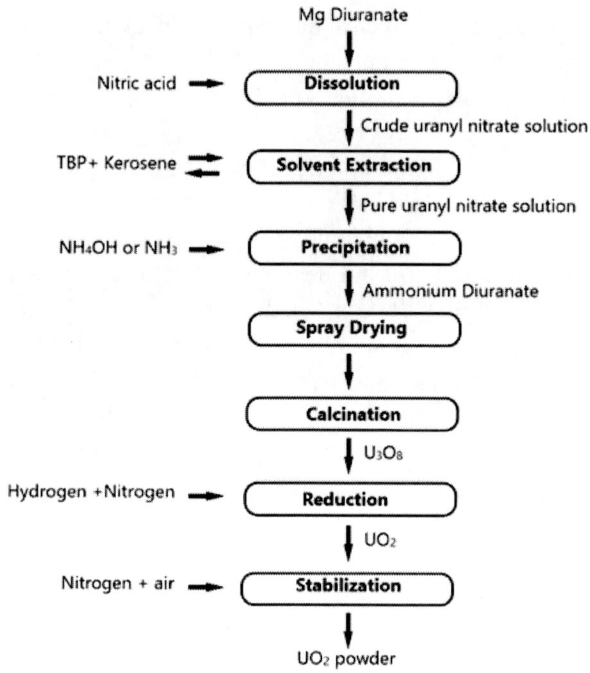

Source: With kind permission from International Atomic Energy Agency.

Figure 11. Flow sheet from magnesium diuranate to UO_2 powder by nitrate route [13].

4. Powder Conditioning

Powder conditioning usually involves stabilization, de-agglomeration by milling, precompaction and granulation, and binder/lubricant addition to enhance flow properties. Fine powders are required for sintering compacts to high densities. However, fine powders may tend to agglomerate and hinder flow. The agglomerates may gain strength in thermal treatments such as drying and calcination. The presence of even a minor percentage of agglomerates may cause sporadic low sintered density leading to rejection of the entire lot. Hence milling is carried out to eliminate agglomerates. Hammer milling and attritor milling are widely used in the industry. Fine powders also do not flow readily. This makes homogenization difficult. The powder also does not flow uniformly into the die fill; therefore, the powder needs to be granulated for subsequent processing. The granulation may be achieved by pre-pressing into slugs or roll pressing and breaking them through a sieve. Granules may also be made by spray drying. Then a binder/lubricant agent is added to the powder. A well-conditioned powder flows well and gives green pellets of adequate strength and with reproducible characteristics.

4.1. Stabilization of UO_2 Powder

Fine UO_2 powders are pyrophoric, just as fine metallic powders are. The heat evolved on oxidation of one UO_2 particle catalyzes the oxidation of neighboring particles in storage, thereby triggering a 'chain' reaction of oxidation. The powder is stabilized against oxidation during handling and processing. On exposure to limited air or to carbon dioxide at room or low temperature, the particles get coated with a higher oxide, which then limits access of the particle to oxygen in the air. The protection achieved is similar to the passivity of aluminum or chromium or titanium or zirconium.

Carbon dioxide gas is very mildly oxidizing to UO_2. The product of mild oxidation is a thin layer of higher oxide that forms on the surface of the UO_2 powder particles and stabilizes the powder against further oxidation in the bulk. The process of stabilization rests on two principles (a) restricting the availability of oxygen to the powder, and (b) releasing the heat of oxidation to the atmosphere instead of to neighboring particles, so spreading out the particles helps in stabilization.

4.2. Parameters for Flow Properties of Bulk Solids

Narrower particle size distributions show better flow behavior than broader distributions, even if the mass median of the narrower distribution is somewhat smaller than that of the broader distribution.

A powder flows when the mobilizing gravitational forces are higher than particle-particle interaction forces. The two fundamental particle-particle interaction forces are cohesion and friction. Friction acts tangentially to the contact point surface and opposes the relative motion of a particle. It increases with the true contact area. Cohesion is the mutual attraction between two particles, which acts perpendicularly to the tangent of the contact point surface. It is defined as the resistance to separation of contacting powder particles of identical material.

The total force of adhesion is a summation value of several additive force components.

$$F_{ad} = F_{vdw} + F_c + F_e + F_{es}$$

where F_{ad} is the force of adhesion, F_{vdw} is the Lifshitz-van der Waals' force, F_c is the capillary force, F_e is the electrical force, and F_{es} is the electrostatic (Coulomb) force.

In general, powder flow is influenced by environmental factors, such as moisture and static electricity, as well as powder related factors, such as morphology, size, size distribution, true density, surface area, and chemical composition. Additionally, the previous history of the material, which means the previous consolidation stress, can have an influence on powder flow ability [14].

4.3. Powder Milling

In the as-prepared green compacts, both small and large pores can be found. The small pores shrink and tend to disappear upon sintering, whereas the large pores remain, despite shrinkage, or may even grow due to the pressure of gases trapped within the closed pores. The size of the large pores may be larger than the critical pore size. Therefore, the maximum density that can be achieved is limited by the volume fraction of large pores present initially [15]. The extent of large pores in a green compact is greatly reduced by milling the powder before compaction, due to a reduction in particle sizes. For example, Figure 12 shows that the particle size distribution (PSD) mean has shifted from 11 to 2 μm after milling.

Powders made from UF_6 usually contain hard agglomerates. Hence, the powder is subjected to milling to break down the agglomerates into smaller entities. Traditionally, ball mills have been used in laboratory work, and hammer mills have been used in tonnage plants. The attritor has taken the place of the hammer mill in some nuclear fuel industries, especially in mixed oxide fuel (MOX). A small screen size in the mill means a smaller agglomerate size in the powder.

Figure 13 shows the beneficial effect of light attrition milling of 5 minutes on green density of UO_2. The powder was prepared by precipitation of ADU from uranyl nitrate.

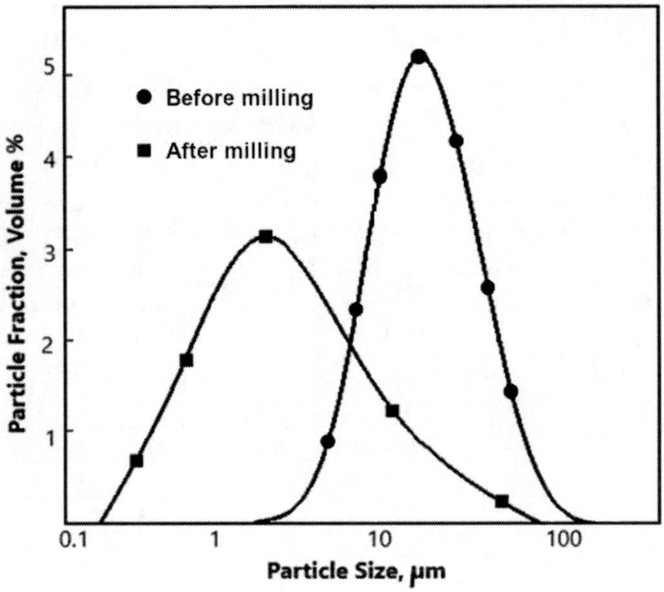

Figure 12. The PSD mean has shifted from ~11 μm before milling to ~2 μm after milling [16].

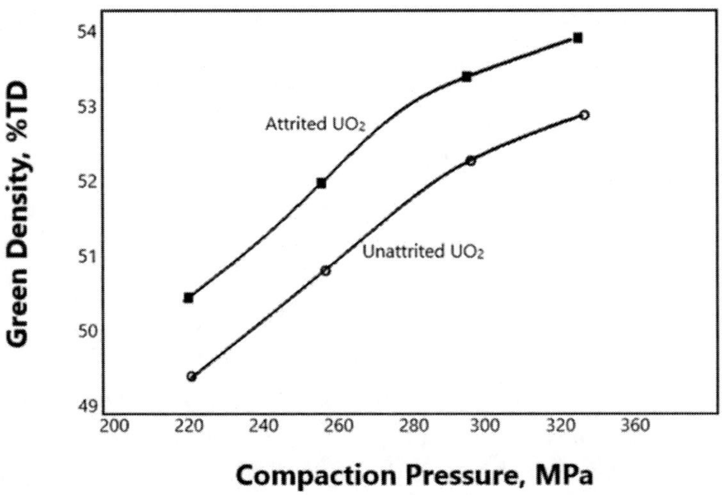

Figure 13. Effect of light attrition milling of 5 minutes on green density of 10 kg batches of UO_2 – the upper line is for pellets from attrited powder [17].

4.4. Granulation

Sinterability (ceramic activity) and flow properties are opposing requirements in UO_2 powder, and usually sinterable powders flow poorly. Hence, most UO_2 powders go through granulation, a process that imparts flowability to a nonflowable powder. This has been achieved by pressing the powder into large slugs, breaking them up, and passing them through a 14 or 20 mesh screen.

A roll press may be used in place of a precompaction press. Spray driers have also been used to obtain flowable granules from uranyl nitrate.

Table 2. Comparison of agglomerates and granules

	Agglomerates	Granules
1	Formation unintentional	Intentionally made
2	Hinder powder flow	Flow well
3	Irregular shaped	Rounded
4	Wide size range	Narrow size range
5	Inter particle bond strong	Inter particle bond weak
5	Solid bridges due to thermal history	Van der Waal bridges, no post thermal treatment before final pressing, hence no solid bridges
6	Retain individual identity in pressing and sintering	Individual identity is lost in pressing and sintering
7	Formation is affected by precipitation and calcination parameters	Granule strength is affected by slugging pressure.
8	Deagglomeration by milling	Granulation by pre-slugging, roll compaction or by spray drying

4.5. Binders or Lubricants

The powder granules are usually coated with zinc stearate or Acrawax (N, N'-ethylene bisstearramide) or a mixture of stearic acid and palmitic acid powder to bring down compaction and ejection pressures and to increase packing efficiency in ceramic powder processing. The stronger and denser compacts obtained through these methods can then be sintered to achieve high densities with minimal defects and shrinkage, leading to high acceptance rates in large scale production. While zinc behenate has also been used as an admixed binder [14], the sintering cycle must be carefully adjusted to ensure complete binder burnout before the material reaches the high-temperature zone to prevent residual carbon contamination in the final product. However, the use of zinc compounds can result in metallic deposits in the cooler regions of the furnace and are not usually preferred.

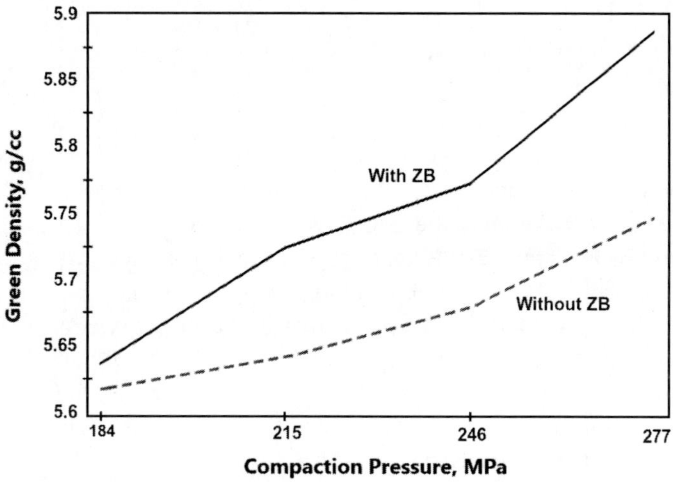

Figure 14. Green Density vs Compaction Pressure. Upper curve is for UO_2 admixed with 1 wt.% zinc behenate. The lower curve is for UO_2 without ZB [18].

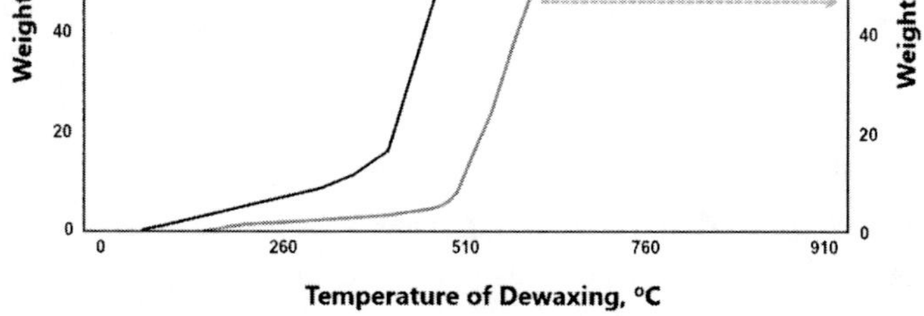

Figure 15. Weight loss % of lubricant versus temperature. Upper curve is for free lubricant. The lower curve is for UO_2 pellet with 1 wt.% zinc behenate [18].

Figure 14 shows evaporation behavior of zinc behenate as a free lubricant and as mixed in UO_2. Figure 15 shows green density versus compaction pressure for UO_2 with and without zinc behenate additive. For the same compaction pressure, a higher green density is obtained when zinc behenate is used.

5. Pressing

When loose powder (that has not been subjected to pressure) is sintered, a low-density porous mass is obtained. On the other hand, when a high sintered density is required, the powder must be compacted to a certain green density. The compaction process provides the required shape and the inter-particle contacts needed in sintering. 40 to 50% of the volume of a green compact may be vacant space. In sintered UO_2, some residual porosity is needed for swelling resistance under irradiation. The sintered density may be ~ 95% TD. Usually, the higher the green density, the higher the sintered density. In axial pressing, the compaction pressure is not transmitted equally throughout the powder mass. There is loss of pressure due to inter-particle friction and die wall friction. Hence, the green density is not uniform throughout the pressed compact. Isostatic pressing is expected to yield a more uniform green density, but is not economical for nuclear fuel production. Grinding allowance is provided in the sintered pellet to be able to remove differences in diameter caused by shrinkages from different green densities in the sintered body. Sintering shrinkage allowance and green pellet expansion allowance are provided while determining the press die bore diameter.

5.1. UO_2 Powder to Pellet

Figure 16 is an example of a flow sheet from UO_2 powder to pellet.

5.2 The Significance of Green Density in Sintering

When loose powder is passed through a sintering furnace, there is no significant densification, though there is some strengthening. Loose powder sintering may be used for making porous materials such as metallic or ceramic filters. When substantial densification is needed, as in the case of dense structural ceramics and nuclear fuel, the powder is pressed into compacts of a certain density before sintering. In addition to providing the required near-net shape, the pressing operation provides the much needed inter-particle contact and increases the particle coordination number. The green compact density is usually of the order of 50 to 60% TD, and the compact is permeable to fluids due to its open porosity. The contact between particles in the green compact facilitates densification in sintering by diffusional transfer of matter into vacant spaces.

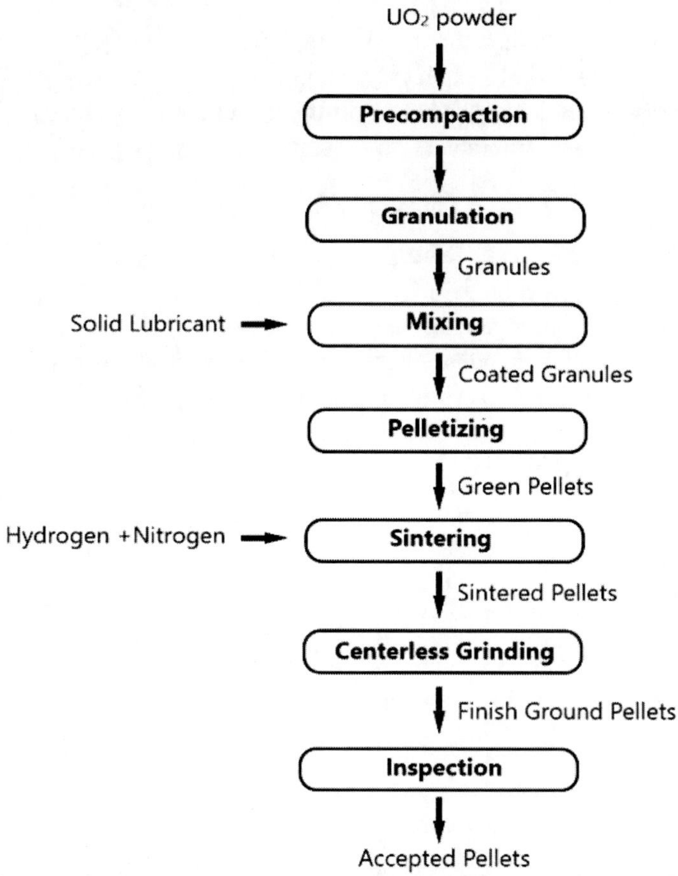

Source: With kind permission from International Atomic Energy Agency.

Figure 16. Flow sheet from UO_2 powder to pellet [13].

Mechanically, the process consists of imposing a progressive strain on the powder, confining it to a certain final volume and porosity. The dimensional constraints imposed by the punches and die are then removed, and the compact is allowed to relax. The compacted material exerts a certain stress against the punches and die during this process in response to the imposed strain.

In the pressing process, the green density of the compact increases as the pressing pressure is increased, as shown in Figures 17 and 18.

Green density plotted against pressing pressure gives an idea of the nature of the powder. Soft powders result in higher green densities at relatively low pressing pressures. On the other hand, hard powders require higher pressures. Agglomerated powders are expected to show lower green density for the same pressing pressure and display breaks in the curve corresponding to fracture events of powder agglomerates.

5.3. Porosity in the Green Compact

When a powder is homogeneous and free from agglomerates, then the green compact is also homogeneous. The pores are uniformly fine in size, and the green density distribution across the compact is also uniform. In such cases, full densification (near 100% TD) is achieved when the green density exceeds a threshold of about 50% TD.

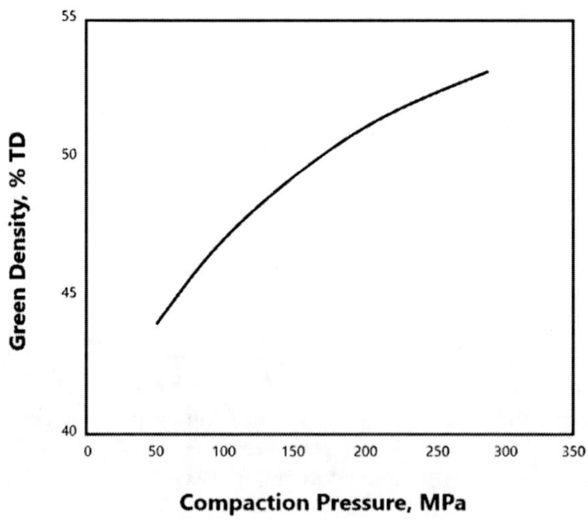

Figure 17. Green density of UO_2 versus pressing pressure [19].

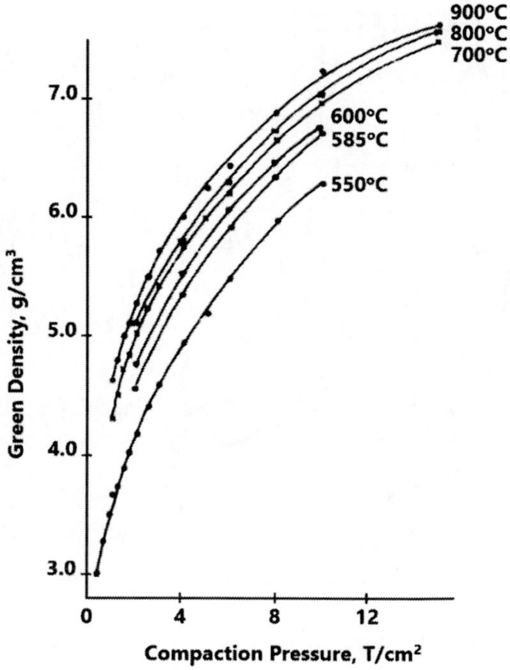

Figure 18. Green density of UO_2 versus Pressing pressure, for different reduction temperatures [10].

The green density of a pellet is of the order of 50 – 60% TD. This means that the porosity in the green pellet is 40% to 50% by volume. The average pore size in the green compact depends on green density and average particle size.

$$d_{pore} = 2\ D_{particle}\ (1\text{-}f)/\ 3f$$

where, d_{pore} is the average diameter of the pore, $D_{particle}$ is the average diameter of the particles, f is the fractional density. Assuming a fractional density of 0.6,

$$d_{pore} = 0.27\ D_{particle}$$

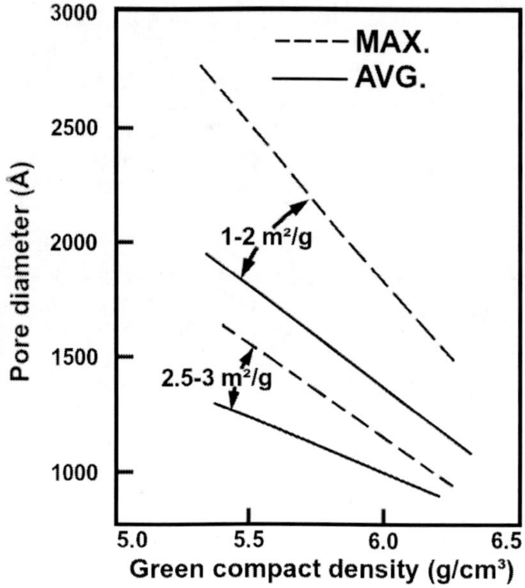

Figure 19. Pore diameter in UO_2 green compacts as a function of green density [20].

The above expression shows how the pore diameter increases as the particle diameter is increased. Large pores in the green body remain after sintering. The strength of the sintered pellet deteriorates when large pores are present in the sintered body.

Figure 19 shows that the pore sizes in the green pellet are reduced as the compaction pressure is increased. It also shows that pores are smaller for higher BET powders.

5.4. Density Gradients in a Green Pellet

Local variations in density are inherent in a green pellet made by axial pressing, as shown in Figure 20. In isostatically pressed material, the density gradients are expected to be less than in axial pressing. The green density of the pellet versus compaction pressure was obtained by experiment. Compaction pressure is estimated in different regions of a pellet model and corresponding green density from experiment is assigned to the region. The r-coordinate is radial, with zero at the centerline of the pellet; z-coordinate is along the pressing axis, with zero at the mid plane of the pellet [19].

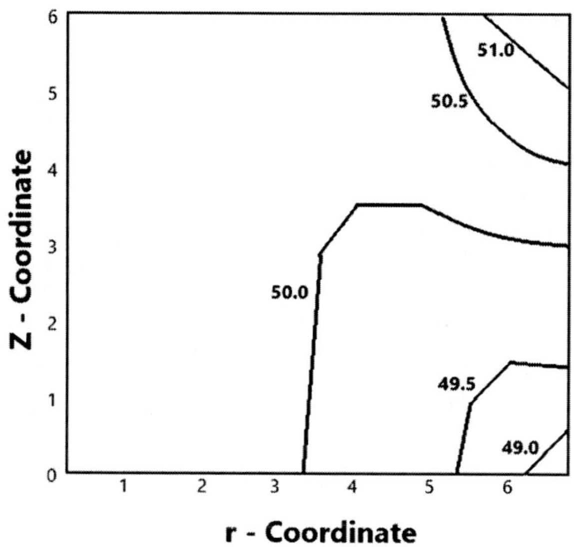

Figure 20. Density contours in a pressed UO_2 green pellet (double acting press, flat pellet) [19].

In contrast to uniaxial pressing, isotactic pressing utilizes a fluid medium to transmit pressure more uniformly to the powder, leading to a more homogeneous compaction.

5.5. Estimating Die Bore Diameter

If the mass loss (of ~1%) in the pellet on sintering is neglected and if the height of the pellet is taken to be the same as that of the diameter, green and sintered densities are related by

$$d_{sint} = d_{gr} \left(\rho_{green} / \rho_{sint}\right)^{1/3}$$

where d_{sint} = diameter of the sintered pellet, d_{green} = diameter of the green pellet, ρ_{green} = density of the green pellet, ρ_{sint} = density of the sintered pellet. This expression is useful in estimating the green diameter that needs to be maintained to prevent unground patches for UO_2 powder blends of known sintered density.

The diameter of the ejected green pellet is slightly larger than the pressing die bore diameter.

Green pellet diameter is given by

$$d_{green} = (1+f_e)\, d_{bore}$$

where f_e is the fractional increase in diameter of the green pellet on ejection from the die and d_{bore} is the diameter of the die bore. f_e is determined by experiment and depends on the pressing pressure and die material. It is usually greater than zero. Therefore, one can never reinsert the pellet into the die bore. The extent of expansion depends on the value of the green density and the rigidity of the die material.

On sintering, the pellet densifies, and the pellet diameter shrinks. Sintered pellet diameter is given by

$$d_{sint} = (1-f_s)\, d_{green}$$

where f_s is the fractional shrinkage in diameter, determined by experiment.

The diameter is reduced further in centerless grinding. Ground pellet diameter is given by

$$d_{ground} = d_{sint} - \Delta d$$

where $_{\Delta}$d is the grinding allowance.

Combining the above expressions and rearranging the terms, the die bore diameter can be calculated:

$$d_{bore} = [d_{ground} + \Delta d]/[(1-f_s)\,(1+f_e)]$$

6. Sintering

Just as liquid drops or soap bubbles in contact can coalesce together, solid particles too can merge into a single entity. Capillary forces or forces of curvature are responsible for the coalescence. By forming a common boundary, the overall surface energy is reduced. While liquids can coalesce at room temperature, solids require high temperatures for the movement of atoms or ions. This is achieved in the process of sintering, by heating the powder compact to a temperature below the melting point. The pressing operation provides the shape needed and numerous inter-particle contacts through which matter can diffuse during sintering. The density and strength of the sintered body are considerably higher than that of the green body. Dimensional shrinkage is proportional to the increase in density. In a two-dimensional visualization of sintering, particles represented by circles change into hexagons. Curved surfaces become straight and the vacant spaces between the circles disappear on sintering. In three dimensions, spheres in contact transform into truncated octahedrons to minimize surface area. The sintering process is useful when shaping materials of very high melting points where melting and casting are not economical, for example molybdenum, tungsten, and UO_2. It is also the only technique available when there is a large difference in melting points, as in the case of tungsten-copper electrical contacts. Sintered density and shrinkage increase as the sintering temperature increases. For the same sintering temperature, sintered density increases as the green density increases. Residual porosity may persist under some conditions.

6.1. Driving Force for Sintering

Reduction in surface energy provides the driving force for sintering. The powder particles need to be very fine and remain in intimate contact with one another to facilitate atom transfer later under the influence of heat. The intimate inter-particle contact is achieved by pressing. In other words, the coordination number of each particle has increased above that prevailing in loose powder. The coordination number of a particle here refers to the number of other particles with which it is in direct contact.

The temperature must be high enough for the compact to make the atoms sufficiently mobile to move into vacant spaces. This is achieved in sintering. On the other hand, if loose powder is sintered without pressing, there is some shrinkage, but the inter-particle porosity mostly remains in the sintered body. Loose powder sintering is one way to make filters and porous pellets. Another way to make porous pellets is to compact a higher oxide and then sinter in a reducing atmosphere. U_3O_8 powder has been pressed and sintered to make porous UO_2 pellets [21].

The large surface energy of small particles provides the driving force for the sintering of a powder compact. The particles join at points of contact, forming necks. The growth of these necks is uninhibited in a powder of a glass because the total surface energy continues to decrease as they get larger. In a crystalline powder, however, the formation of a neck invariably generates a grain boundary. Neck growth proceeds only when the driving force for sintering, provided by the decrease in total surface energy, is greater than the energy cost of grain boundary formation. This generally facilitates neck growth, because the grain boundary energy of a solid is typically about half as large as the surface energy. Grain boundaries serve as sources of matter to fill voids between the particles [22].

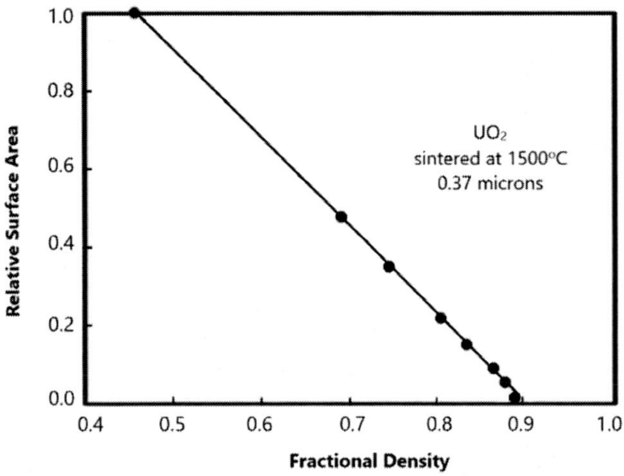

Figure 21. Surface area (normalized to the starting powder surface area) versus fractional density for urania sintering at 1500°C [24].

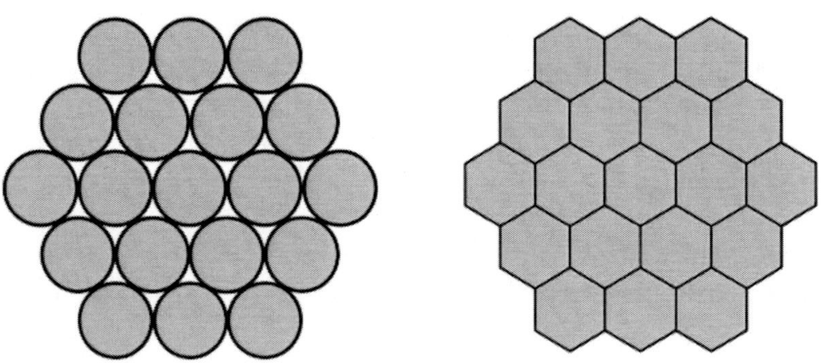

Figure 22. Two-dimensional representation of particles in contact. Sintering of circles leads to formation of hexagons accompanied by curve straightening and shrinkage. Even though all the circular particles have been shown to have transformed into hexagonal grains, the larger grains grow at the cost of the smaller grains.

Figure 21 illustrates the drop in Specific surface area versus fractional density of UO_2 undergoing sintering.

In two dimensions, particles are represented by circles. Circles in contact transform to hexagons on sintering, as illustrated in Figure 22. The circles in contact (shown on the left) have two features – curvature and voids. The hexagons in contact (shown on the right) have lost both the curvature and the voids. Even though all the circular particles have been shown to have transformed into hexagonal grains, the larger grains grow at the cost of the smaller grains.

In a three-dimensional representation, the particles are viewed as spheres in the green compact which change to truncated octahedrons on sintering.

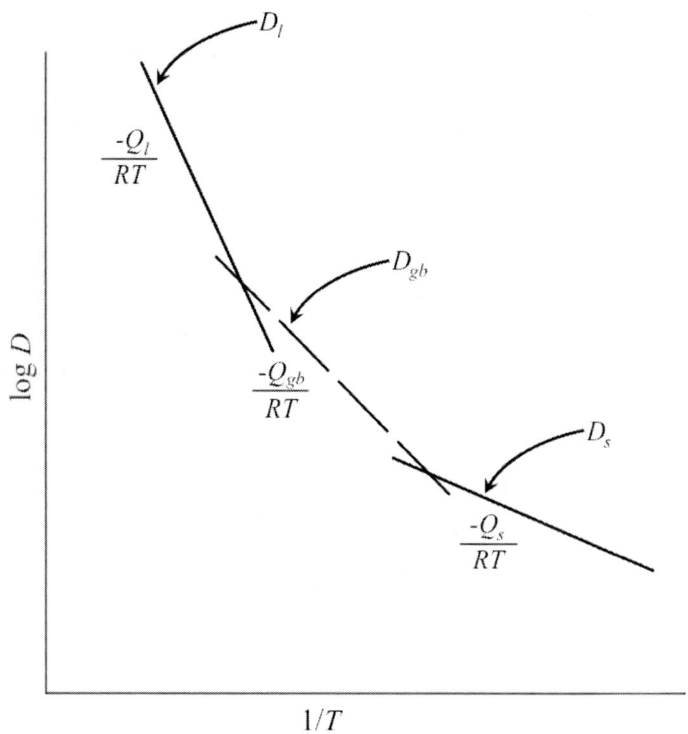

Figure 23. Dominant mechanisms of diffusion in different temperature ranges [23]. Note that the temperature decreases along the x-axis.

The plot in Figure 23 depicts the dominant mechanisms operating in different temperature ranges, where the x-axis represents reciprocal temperature and the y-axis indicates the diffusion coefficient. Surface diffusion is dominant at low temperatures (favoring particle coarsening), grain boundary diffusion is dominant at intermediate temperatures, and lattice diffusion is dominant at high temperatures (favoring densification). This information may be useful in designing furnace temperature profile so that densification is promoted, and particle coarsening is limited.

6.2. The Effect of Calcination/Reduction Temperature of Powder on Sintering

The sintering behavior of a powder compact is dependent on its thermal history. Figure 24 is a dilatometer output describing the sintering behavior of UO_2 powders reduced at temperatures varying from 630°C to 900°C. There is a decrease in BET specific surface area of the powder as the reduction temperature is increased. There is also a corresponding shift in the shrinkage behavior in sintering for higher temperature reduced powders.

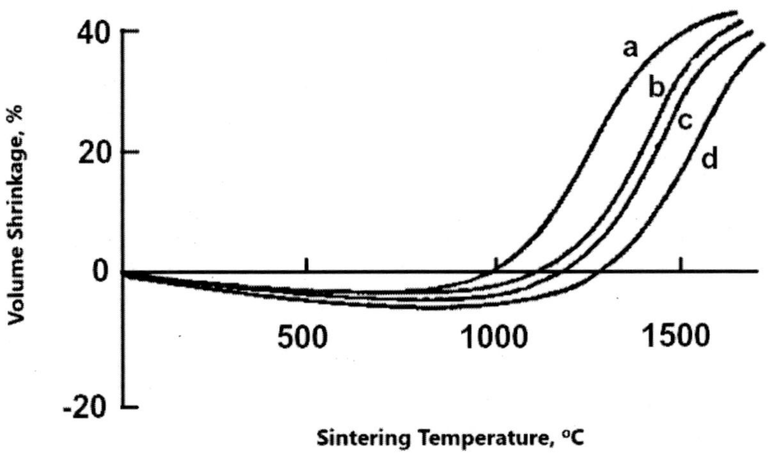

Figure 24. Volume shrinkage versus temperature of UO_2 pellet in a heating microscope [10]. Initially, there is some thermal expansion. Heating rate: 390°C/hr. a: 5.7 m^2/g (630°C), b: 4.4 m^2/g (700°C), c: 3.4 m^2/g (800°C), d: 2.0 m^2/g (900°C). Note the effect of reduction temperature on the onset of shrinkage. For the same shrinkage, higher sintering temperatures are needed for pellets from powders that have seen higher reduction temperatures.

6.3. Green Density, Temperature, and Sintered Density Relationship

Figure 25 shows the increase in sintered density versus sintering temperature in ZnO compacts. In this case, different green densities have resulted in the same sintered density. This powder is evidently free from agglomerates and the final sintered density is the same, whatever the green density is after the threshold green density is crossed. A similar situation may be applicable to UO_2.

When the same sintered density is achieved for different green densities, it means that the sintering shrinkages are different. A low green density compact shrinks much more than a high green density compact does. Pressing die diameters are kept larger than the sintered pellet diameter considering the shrinkage expected on sintering.

Low green density is easy on tooling but tends to produce a larger number of defective sintered pellets because the green pellets are more fragile. The higher green densities are harder on the tooling and can produce capping and other negative effects due to excessive green pellet rebound.

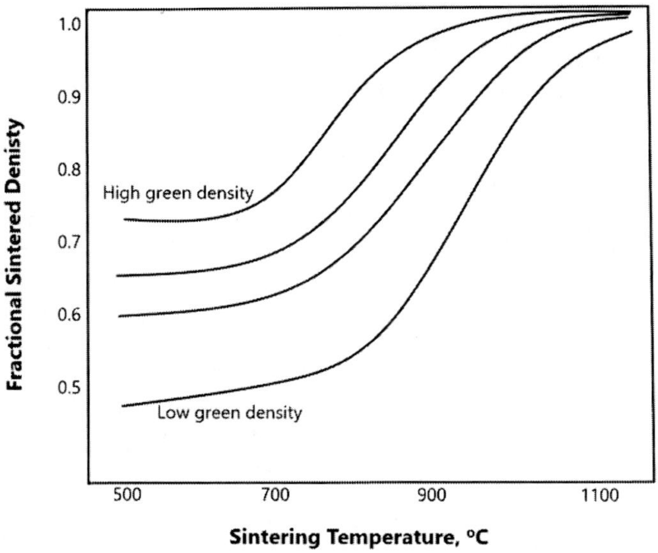

Figure 25. Effect of green density on the sintering of ZnO powder compacts during constant heating rate (5°C/ min) to 1100°C showing that for this powder, the final density is almost independent of the green density for a wide range of green density values [25].

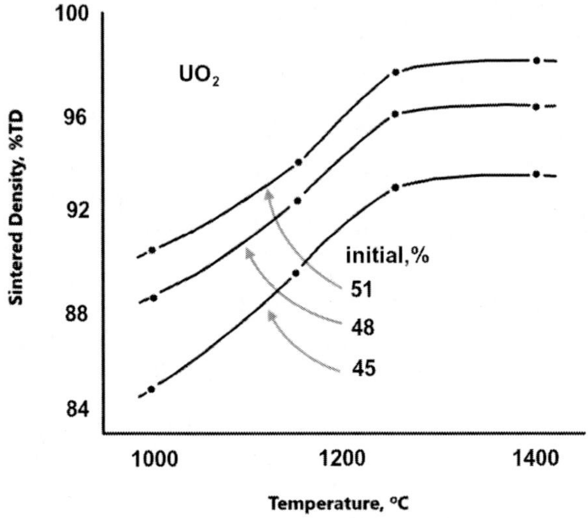

Figure 26. Sintered density versus sintering temperature for UO_2 [24].

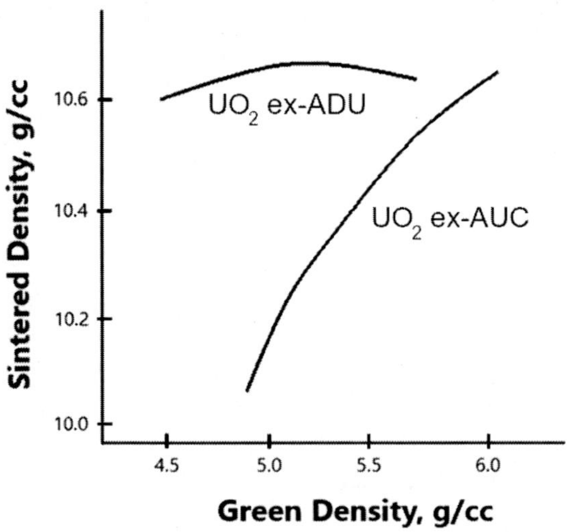

Figure 27. Sintered density versus green density for AUC and ADU derived pellets [1].

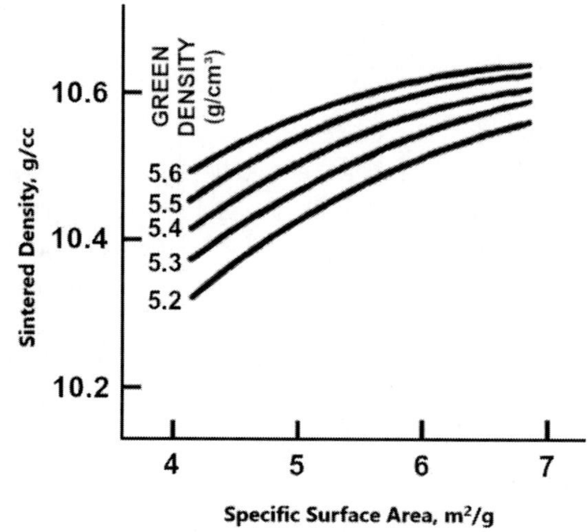

Source: With kind permission from International Atomic Energy Agency.

Figure 28. Sintered density vs. surface area for different green densities The UO_2 powder was from AUC [26].

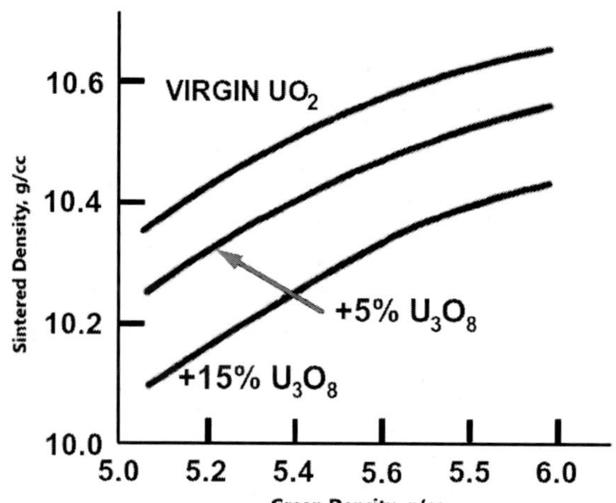

Source: With kind permission from International Atomic Energy Agency.

Figure 29. Sintered density vs. green density for different U_3O_8 contents. **[26]**. U_3O_8 is added to utilize pure scrap UO_2 and to depress sintered density. The UO_2 powder was from AUC.

The variation of sintered density of UO_2 with increasing temperature is shown in Figure 26. The behavior of UO_2 is seen to be different from that of

ZnO shown in the previous figure in that the final sintered density is more strongly dependent on the green density.

Figure 27 shows the sintered density of UO_2 as a function of green density. This figure highlights the difference in the sintered density relationship based on the UO_2 source. For UO_2 derived from ADU, the sintered density remains approximately constant regardless of the green density. In contrast, UO_2 from AUC exhibits a greater sensitivity to variations in green density, at least above 4.5 g/cc. Figure 28 illustrates the variation of sintered density with powder specific surface area for UO_2 produced via the AUC route. The sintered density increases with increasing specific surface area, with the curves plotted for different green densities. Figure 29 depicts the sintered density as a function of green density, along with the effect of U_3O_8 addition. The addition of U_3O_8 results in a lower sintered density.

6.4. Porosity Effects

Even minor variations in the original particle packing are exaggerated during the pore growth process. Spaces between agglomerates and occasional larger voids resulting from the bridging of particles or agglomerates may be present. As a result, during intermediate stages of the sintering process there is a range of pore sizes present. The slower elimination of the larger pores leads to variations in pore concentration in the later stages of the sintering process.

The importance of local variations in pore concentration results from the fact that the part of the sample containing pores tends to shrink but is restrained by other pore-free parts. That is, the effective diffusion distance is no longer from the pore to an adjacent grain boundary but from a pore-pore or pore-surface distance many orders of magnitude larger.

6.5. Pore Shape, Size, and Stability

The equilibrium shape for pores within a grain is a sphere with a minimum surface area for the volume. On the other hand, pores lying at grain junctions and pores intersecting grain boundaries adopt a shape that is not a sphere. The shape is determined by the balance of interfacial energies between the grains and the pore.

Pores within a grain cannot easily grow or shrink as the access to material transfer is through lattice diffusion, a slower process compared to grain boundary diffusion. On the other hand, pores at grain junctions have a greater chance of material transfer enabling shrinkage or growth. Pores aggregate as grains grow. However, during sintering, there will be an overall and substantial reduction in pore volume.

It is known from the concept of stable density fuel that pores less than about 2 μm in diameter are unstable and close during irradiation. Figure 30. Pores which are larger in diameter, e.g., 20 or 30 μm, remain stable during irradiation, thereby minimizing the problems associated with in-reactor fuel densification.

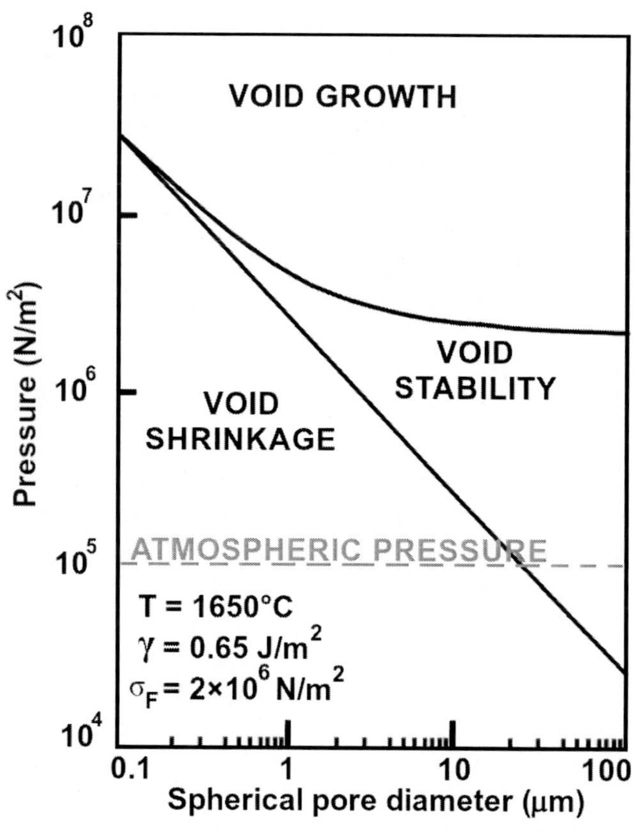

Figure 30. Criteria for void stability in UO_2 pellets sintered at 1650°C [27].

Carefully sized pore formers are sometimes intentionally added to UO_2 powder before pressing so that they would leave pores of a stable size in sintered UO_2. U_3O_8 also behaves like a pore former to accommodate the difference in specific volume between U_3O_8 (TD 8.3 g/cc) and UO_2 (TD 10.96 g/cc). The U_3O_8 agglomerate gets reduced to UO_2 in the preheat zone of the sintering furnace. The phase change leaves a void due to the change in specific volume from that of U_3O_8 to that of UO_2. The size of the pores that U_3O_8 would leave upon reduction in the sintered UO_2 is thus dependent on the starting U_3O_8 agglomerate size. If the U_3O_8 particles are large, they would leave stable pores leading to low sintered densities. If these are small, they would leave unstable pores that would close, leading to relatively higher sintered densities. The closing pressure acting on a pore is represented by

$$p = 2\gamma / r$$

where p is the pressure acting to close the pore, γ is surface energy and r is the radius of the pore. p increases, as r is decreased.

During densification in the sintering furnace, the open porosity is lost when the density reaches ~ 92% TD and there is only closed porosity. If the void size is large enough, the tendency to shrink decreases. In addition, gases from the furnace atmosphere trapped in the large voids oppose the shrinkage force. Pore shrinkage eventually comes to a halt when the two forces balance.

6.6. Densification and Coarsening

Sintering is a process in which powder compacts densify without melting by applying thermal energy. The driving force of this process is the reduction of the total interfacial energy of the system, a phenomenon achieved by coarsening - reducing the solid/solid interfacial area - and densification - reducing the solid/gas interfacial area.

When spheres in contact are heated, the following possibilities may be imagined:

a. Only coarsening, that is, a reduction in specific surface area without a corresponding increase in density. This is represented by the bottom right side of Figure 31.
b. Only densification, that is, an increase in density without significant reduction in specific surface area. This is represented by the top right side in Figure 31.
c. Simultaneous densification and coarsening. This is shown by the diagonally opposite clusters on top left and bottom right.

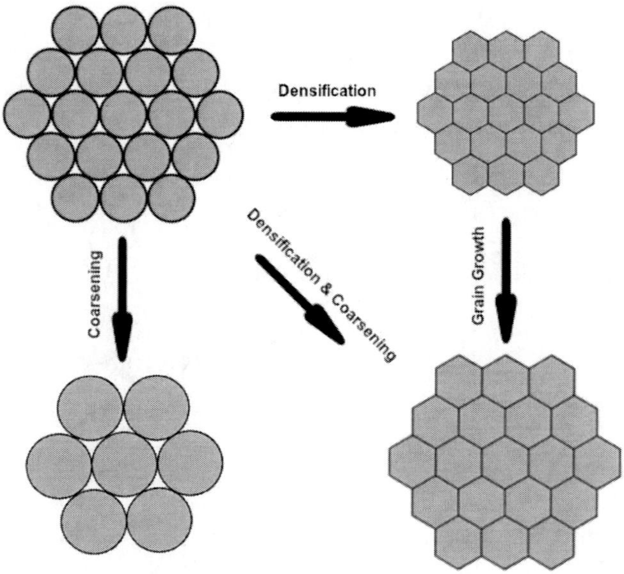

Figure 31. Two-dimensional representation of the possibilities of densification and coarsening in sintering [28].

By coarsening, we mean that the particles have grown, and the pores existing in the green body have been retained and stabilized. Grain growth continues without particle centers nearing each other. Such a situation prevails in the sintering of metallic and ceramic filters that need open porosity.

Packing of particles in a green body is best achieved by using a mixture of different particle sizes, so that the finer particles would occupy the interstices between the larger particles. However, it was realized that the initial physical microstructural inhomogeneity persisted after sintering. At least in advanced technical ceramics where high sintered densities and uniform grain sizes are required, the initial particle size distribution needs to be mono-sized or narrow. This would give a uniform green microstructure that would translate to a uniform sintered microstructure. Figure 32 shows the effect of particle size distribution and sintering temperature on the sintered microstructure. A narrow particle size distribution and a moderate sintering temperature that permits pores to be linked to grain boundaries yields a sintered microstructure with fine grains and free from pores.

A systematic discussion of the conditions for the occurrence of discontinuous growth and the steps to be taken to avoid it was given by Brook [30]. Figure 33 shows the corresponding pore and grain size effect on the growth kinetics for a pure and doped material for the case where the pores move by surface diffusion (if the pores move by a different mechanism, different diagrams result). This figure can be used to see what happens during sintering if we know the sintering trajectory.

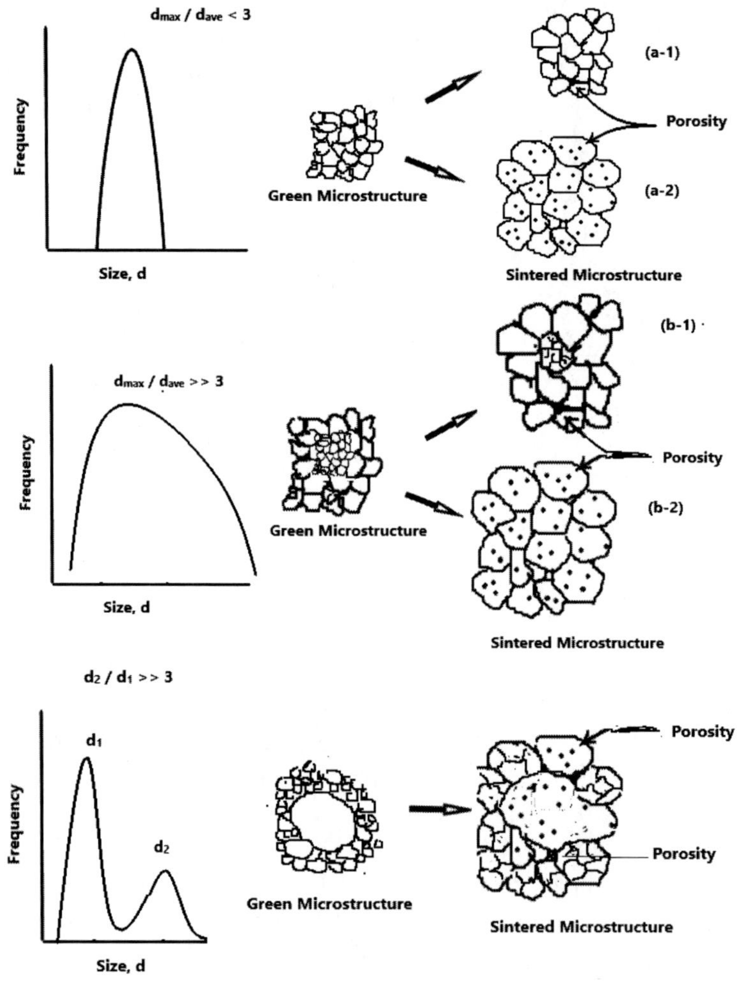

Figure 32. Schematic representation of the effect of powder size distribution on the sintered microstructure. For example, in (a-1), with a narrow size distribution and control of the grain boundary-pore interaction, dense sintered ceramics with the grain size slightly larger than the particle size are possible [29].

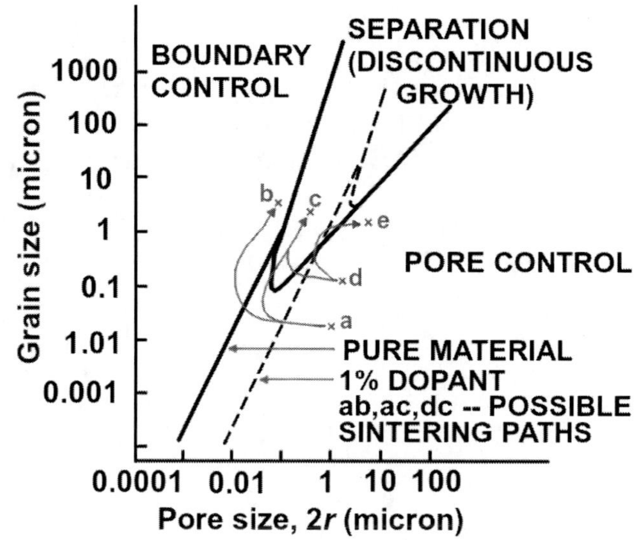

Figure 33. Dependence of the pore-boundary interaction on microstructural parameters for pure or doped material for the case where the pores move by surface diffusion [30].

Three possible paths are shown as *ab, ac,* and *dc.* They begin horizontally to the left due to the dominance of pore shrinkage initially but change to diagonal left bottom to top right because of grain growth. The path *ab* avoids the separation region, and this will lead to a final high-density state; *ac* and *dc,* however, end in the separation region for pure material and will lead to the trapping of pores by cannibal growth. The boundary mobility can be decreased by adding a dopant that is selectively absorbed at the grain boundaries. According to Figure 33, 1% of such a dopant has the effect that trajectories ac and dc end outside the separation region so that a high density may be attained.

An admixed binder/lubricant agent used in pressing is required to leave the green pellet in the low temperature zone of the sintering furnace. If the binder evaporation is incomplete, it can decompose and leave carbon residue in the sintered pellet, as the furnace atmosphere is reducing in nature. The presence of residual carbon may, in turn, affect other impurities. UO_2 can react with carbon in the presence of nitrogen and increase the nitrogen content in the sintered pellet. The gas flow pattern in the sintering furnace must be adjusted so that the evaporating binder/lubricant agent is swept away by counter current gas flow, so that there is no chance of carbon residue in the higher temperature zones. Figure 34 shows counter current flow that is conducive to binder removal. Figure 35 shows co-current flow that is conducive to nitridation of UO_2.

The presence of moisture in the reducing atmosphere of the sintering furnace is considered beneficial for densification [32]. On the other hand, if moisture migrates to the cooling zone, it can crack or oxidize the pellets in the cooling zone.

Figure 36 shows an example of pellets in a boat. Figure 37 shows an example of pellet loading pattern.

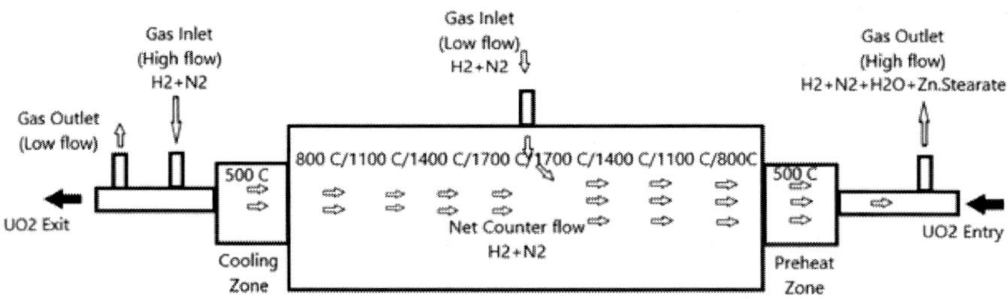

Figure 34. The input gas flow rates into the UO_2 sintering furnace are such that there is a net counter-current flow, conducive to removal of carbonaceous matter and moisture [31].

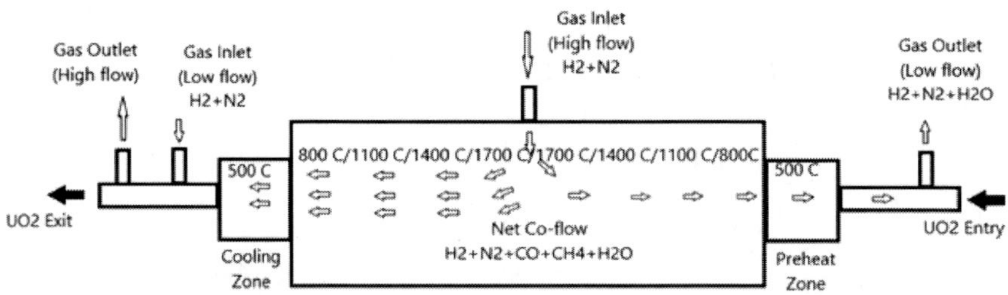

Figure 35. Here, the input flow rates into UO_2 sintering furnace are such that there is a net co-flow, conducive to nitridation and oxidation [31].

Source: With kind permission from International Atomic Energy Agency.

Figure 36. An example of boat with pellets [33].

Source: With kind permission from International Atomic Energy Agency.

Figure 37. An example of pellet loading pattern [33].

7. Grain Growth and Pore Stability in Sintering

Both densification and grain growth take place during sintering. As in the case of densification, grain growth too increases with temperature and time. Some applications (transparent Al_2O_3) require the grains to be small. Other applications require the grains to be large (UO_2 pellets). The grain growth rates can be controlled by temperature, additives, and manipulating sintering parameters. Larger grains grow at the cost of smaller grains. Grains with a smaller number of sides or faces tend to shrink, while those with a larger number remain stable or even tend to grow. Additives may serve as grain growth inhibitors or grain growth promoters. Premature grain growth may inhibit elimination of residual porosity. The stability of a void depends on its size, the dihedral angle, and the number of grains intersecting the void. Pores linked to grain boundaries are found to shrink more readily than isolated pores. Pore growth may happen because of grain growth.

7.1. Driving Force for Grain Growth

The grain boundary is a surface enveloping the crystal. Grain growth is a natural process where the total grain boundary area decreases. The average grain size increases and the total number of grains in the material decreases. It is the tendency of nature to reduce surface energy, including grain boundary energy. Grain growth is driven by the reduction in grain boundary energy. However, an activation energy barrier must be overcome for grain growth. When subjected to heat, ionic mobility increases, and the ions can move from one grain to a neighboring grain. This leads to grain growth. The rate of grain growth increases as the temperature is increased. The growth of one grain is accompanied by the shrinkage of another grain. Grains grow at the cost of smaller grains that eventually disappear. Energy is reduced in the process of grain growth. Some additives inhibit grain growth whereas other additives promote grain growth.

The grain growth in a sintered ceramic is dependent on temperature and time. The higher the temperature and the longer the time, the greater the grain size reached. The dependence of grain size on temperature, time and initial grain size has been expressed as below:

$$G_t^m - G_0^m = K\,t\,e^{(-Q/RT)}$$

G_0 = initial grain size
G_t = grain size at time **t**
T = absolute temperature
Q = activation energy
m = an exponent
R = Universal Gas Constant.

General width of a grain boundary is of the order of 5 Angstroms. Atoms from the convex side of the grain move to the concave side of the grain surface. The resulting atomic flux induces the boundary to move toward its center of curvature. Chemical potential difference across the two surfaces is responsible.

The sintering of copper wires is illustrated in Figure 38 at three different temperatures [34]. Initially, the sintering rates are higher for the higher temperatures. However, the final density reached was higher for the lower sintering temperature. This has been explained based on pores linked to grain boundaries. If the pores intersect grain boundaries, vacancies can migrate out of the pore via the grain boundary. This means that atoms can migrate into the pores via grain boundaries. Once the grain growth is rapid, pores may get detached from the grain boundary thereby losing one avenue for mass transfer. A similar mechanism has been assumed to be operative in the sintering of UO_2 [35].

The use of additives may promote densification or restrain grain growth. The sintering temperature profile may also be moderated to restrict grain growth. The sintering atmosphere may be modified to favor grain growth.

MgO has been shown to inhibit grain growth in the sintering of Al_2O_3 and ThO_2. Inhibition of grain growth was found to enhance densification as grain boundaries intersecting pores act as avenues for transporting vacancies from the pores. TiO_2 and Nb_2O_5 promote grain growth in UO_2. Cr_2O_3 or a combination of Cr_2O_3 and Al_2O_3 with some modifications of the sintering atmosphere have also been used for obtaining large grains.

7.2. Pore Growth due to Grain Growth

Pores can, like grains, grow or shrink and there may be pore agglomeration consequent on grain growth. Pores at intergranular junctions are usually larger than those within the grain due to pore aggregation caused by grain growth.

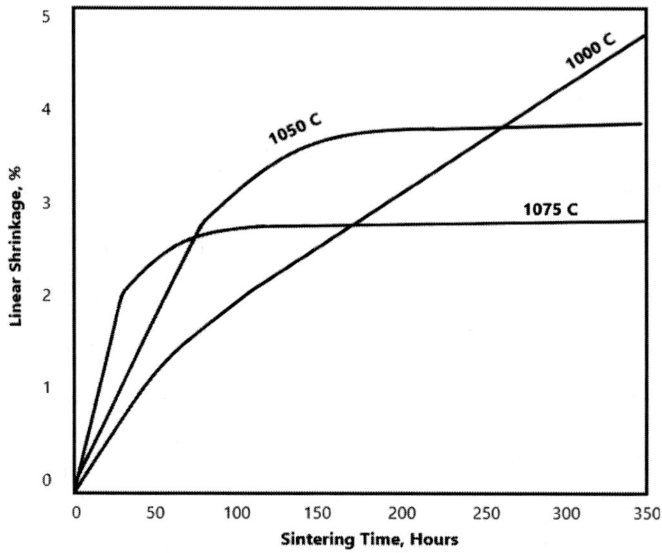

Figure 38. Shrinkage versus time of sintered copper wires. Even though the initial sintering rates are higher for higher temperatures, final density is lower [34]. The phenomenon has been observed in ceramics as well.

7.3. Pore Stability Versus Dihedral Angle and Grain Coordination Number

On the lines of grain growth, pore growth too is affected by curvature. Generally, pores with few neighboring grains tend to shrink. Small pores with few neighboring grains have concave surfaces and tend to shrink, whereas large pores with many neighboring grains have convex surfaces and tend to grow. It is assumed here that there is no pore aggregation caused by grain growth. If the grain growth is fast enough, it can reduce the number of neighboring grains to the pore and help in shrinkage if the grain boundary is connected to the pore.

A curved grain boundary tends to straighten by moving towards its center of curvature. The stability of a pore is governed by the relative size of the pore to the grain size and the relative ratio of surface energy to grain boundary energy. For any real system where the dihedral angle ϕ is smaller than 180°, there exists a critical size of isolated pores relative to grain size that cannot shrink by pressure less sintering (atmospheric pressure sintering).

r is the radius of a circumscribed sphere around a polyhedral pore surrounded by grains. ρ is the curvature of the pore surface. The ratio r/ρ decreases as ρ increases. The shape curvature of the pore surface is indicated by r/ρ.

When r/ρ = 0, the surface is flat. When r/ρ = +ve, the surface is convex when seen from inside the pore. The curvature can be lost by the shrinking of the pore. When r/ρ = -ve, the surface is concave when seen from inside the pore. The curvature can be lost by the expansion of the pore. (See Figure 39.)

In real powder compacts where the dihedral angle is, in general, around 150°, the pores smaller than the average grain satisfy the shrinkage condition while those larger than a few times the grain size cannot shrink. In real systems, large pores are more frequently observed than small pores during sintering. The above discussion on pore stability suggests that, in addition to pore coalescence, the high stability of large pores can be another cause of the presence of large pores in many sintering compacts [28].

For UO_2, φ is reported to average at 92°, the grain coordination number for the pore being about 6. The pores will tend to grow if they intersect 6 or more grains. It should be noted that the dihedral angle for UO_2 is much smaller than those normally found i.e., 152° for Al_2O_3 and 130° for Al_2O_3 + 0.25% MgO.

Consider a material with a particular grain size and pore size. Now if the pore size is decreased further, it will intersect less grain boundaries than before and the pore-grain coordination number will decrease. Once the coordination number falls below the critical value, the pore will shrink. The critical pore size is dependent on the dihedral angle.

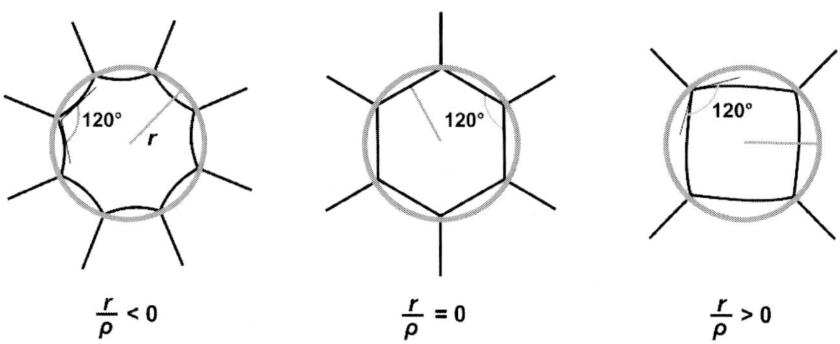

Figure 39. Variation of pore shape with the number of surrounding grains. Dihedral angle assumed to be 120° [36].

Consider two materials, Al_2O_3 and UO_2, with dihedral angles 152° and 92° respectively for a pore surface that is flat ($r/\rho = 0$). If both materials have the same grain size, the critical pore size for Al_2O_3 will be larger than that for UO_2. This means that it can accommodate a larger pore-grain coordination number and hence larger voids that can shrink. These considerations suggest that for the same pore size, the tendency to pore shrinkage is stronger in Al_2O_3 than in UO_2.

7.4. Grain Growth in Uranium Dioxide

Grain growth in UO_2 is affected not only by the type of powder which is used for the fabrication of the sintered pellets, but also by differences in the fabrication procedures applied to the same batch of powder [37].

Grain growth was investigated in UO_2 (94%TD) over the temperature range 1555° to 2440°C [38]. The results were in close, but not exact, agreement with a theoretical expression describing grain growth with a polycrystalline matrix. For the material studied the mean grain diameter D (μm) after annealing for t hours at a temperature T (°K) was given by the equation

$$G_t^2 - G_0^2 = K\, t^{0.8}\, e^{(-87000/RT)}$$

where D_0 and K_0 are, respectively, the initial grain size and a proportionality constant. Uranium metal was found in all specimens annealed above 2000°C. This was taken as evidence that the UO_2 lattice can be oxygen-deficient at high temperatures.

8. Non-Stoichiometry and Point Defect Chemistry

Sintering takes place by diffusion of ions in ionic solids such as UO_2. Both U and O ions must move to fill vacant spaces in the UO_2 green pellet during sintering. The diffusion coefficient of U ions is many orders of magnitude lower than that of O ions. Hence to speed up sintering, U ions must be accelerated. Diffusion depends on point defect concentrations, which in turn depend upon temperature, additive, and atmosphere. When defect concentrations are increased by additives or dopants, it is possible to lower the sintering temperature. Excess O in hyperstoichiometric UO_2 leads to excess U vacancies which speed up diffusion of U ions and permit the use of lower sintering temperatures. The grain growth inhibiting effect of some additives such as Er_2O_3 may be compensated by increasing the oxygen pressure in the otherwise reducing atmosphere of the sintering furnace.

8.1. Non-Stoichiometry in UO_2

The O/U ratio in UO_2 can be less or more than 2 and the deviation increases as the temperature is increased. The O/U depends on the temperature as well as the oxygen pressure in the atmosphere, as shown in Figure 40.

Oxygen can enter or leave UO_2 at will!

- In the UO_2 crystal lattice, a U atom gives up four electrons and becomes a U^{4+} ion. The oxygen atom acquires two electrons and becomes an O^{2-} ion. Overall, there is no net charge on the UO_2.
- When UO_2 is kept under a bell jar and evacuated under heat, some oxygen ions try to leave the crystal, which then becomes UO_{2-x}, sub-stoichiometric.
- Now instead of vacuum, if the bell jar is filled with oxygen gas and heated, then some oxygen penetrates the crystal lattice and become oxygen ions. Then the UO_2 becomes UO_{2+x}, hyperstoichiometric.
- At any given time, both vacancies and interstitials of the same species (e.g., oxygen) are present in the crystal lattice. Only their concentrations change with temperature and environment.

Oxygen inside and outside the crystal

- When the oxygen is in the atmosphere as a gas, there is no net charge. When the oxygen enters the crystal, it acquires two negative charges. Correspondingly, two holes (positive charges) are created.
- Oxygen ions with two electrons like to leave the crystal. Once they leave the crystal, they are no longer charged. Two electrons are left in the crystal. The electrons may remain at the vacant site or move away, becoming part of the conduction band. If this happens, two positive charges are created at the vacant site.
- If UO_2 has negative charges in the conduction band, it will exhibit n-type electrical conductivity.

- If UO_2 has positive charges in the valence band, it will exhibit p-type electrical conductivity. The electronic or hole conductivity increases as the temperature is increased.
- A graph of electronic conductivity versus oxygen pressure is U shaped, especially at high temperatures, as shown in Figure 41.
- The oxygen pressure corresponding to a minimum of electrical conductivity is also that corresponding to the stoichiometric composition of UO_2. It is seen from the graph to be about 10^{-10} atmosphere at normal sintering temperatures. An oxygen pressure above this value is considered oxidative to UO_2 and that below this value is considered reducing to UO_2.

At the sintering temperature, the $UO_{2\pm x}$ can be sub-stoichiometric or hyperstoichiometric depending on the oxygen pressure in the reducing atmosphere of the sintering furnace.

8.2. Mass Transport in Sintering

Both uranium and oxygen ions must move for mass transport to the sintering neck region. This movement is called 'ambipolar diffusion.' The ambi-polar diffusion coefficient is also called the 'effective' diffusion coefficient.

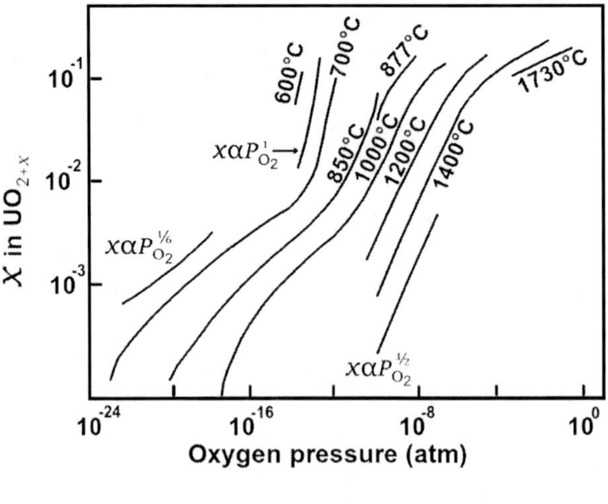

Figure 40. Non-stoichiometry in UO_2 versus oxygen pressure at different temperatures [39].

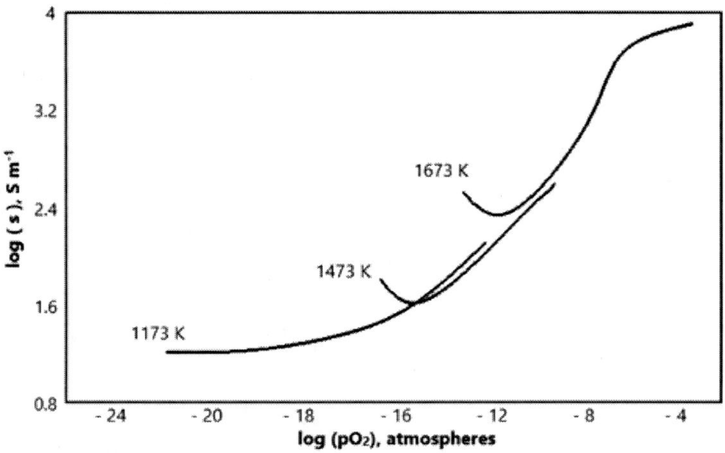

Figure 41. Electrical conductivity of UO_{2+x} versus oxygen pressure [40].

Consider, for simplicity, an MO oxide subjected to an electrochemical potential gradient $d\eta_{MO}/dx$ which in turn must result in the mass transport of MO "units" from one area to another. Typically, this occurs during sintering or creep where, because of curvature or externally imposed pressures, the oxide diffuses down its electrochemical potential gradient. To preserve electro neutrality and mass balance, the fluxes of the M and O ions must be equal and in the same direction.

The driving force here is the chemical potential gradient in MO.

$$D_{MO} = D_M D_O / (D_M + D_O)$$

For the general case of an oxide $M_\kappa O_\beta$, the diffusion coefficient is given by

$$D_{M\kappa O\beta} = D_M D_O / (\beta D_M + \kappa D_O)$$

For UO_2, the ambipolar or effective diffusion coefficient is given by

$$D_{UO2} = D_U D_O / (2 D_U + D_O)$$

This may be rewritten by dividing both numerator and denominator on the left by D_O.

$D_{UO2} = D_U / [2 (D_U/ D_O) + 1]$

However, in UO_2, the oxygen ion diffuses much faster than the uranium ion, 10^6 or 10^7 times.

$D_O >> D_U$

Hence D_U/ D_O may be neglected.

Therefore,

$D_{UO2} = \sim D_U$

In the case of Al_2O_3 sintering, it is the oxygen ion that moves slower than the aluminum ion.

For Al_2O_3, $D_{Al} >> D_O \therefore D_{Al2O3} = \sim D_O$

It is seen from the above that the diffusion rate of the slower moving ion controls densification and grain growth processes.

8.3. Kroger-Vink Notation for Point Defects [39]

- O_O = oxygen ion present in an oxygen site in the lattice.
- O_i = oxygen ion present as an interstitial
- U_i = uranium ion in an interstitial position.
- V_O = oxygen vacancy without charge.
- Dot indicates positive charge, prime negative charge.
- $V_O^{\bullet\bullet}$ = oxygen vacancy with two positive charges.
- U_i'''' = uranium ion with four negative charges.
- $h^{\bullet}$ = hole, a positive charge.
- $Nb^{\bullet}_U$ = Nb occupying a U site, with one positive charge.

A vacancy has zero mass.
Three conditions need to be maintained in writing defect reactions:

- Mass is conserved.

- Electrical neutrality is preserved.
- The number of cation sites is the same as the number of anion sites.

The Schottky defect equation for UO_2 may be written as:

$Null = V_U'''' + 2V_O^{\bullet\bullet}$

The Frenkel defect equation for the cation in UO_2 may be written as:

$U_U^x = U_i^{\bullet\bullet\bullet\bullet} + V_U''''$

For the anion,

$O_O^x = O_i'' + V_O^{\bullet\bullet}$

8.4. The effect of Atmosphere

When oxygen pressure is high,

$\frac{1}{2} O_2 (g) = O_i'' + 2 h^{\bullet}$

which is the same as

$O_2 = 2O_i'' + 4 h^{\bullet}$

UO_2 is now hyperstoichiometric. If the interstitial oxygen manages to get to a lattice site, then a uranium site must fall vacant.

$O_2 (g) = V_U'''' + 2O_O + 4 h^{\bullet}$

Uranium vacancies speed up U diffusion.
When oxygen pressure is low,

$O_O = \frac{1}{2} O_2 (g) + V_O^{\bullet\bullet} + 2 e'$

UO_2 is now substoichiometric.

If the vacant oxygen site manages to obtain an oxygen ion, then a uranium ion must transfer to an interstitial position.

$$2\,O_O = O_2\,(g) + U_i^{\bullet\bullet\bullet\bullet} + 4\,e'$$

Due to Frenkel equilibrium, an increase in the concentration of uranium interstitials means a decrease in the concentration of uranium vacancies and these slow down uranium diffusion.

Figure 42 shows the increase in the diffusion coefficient of uranium ions in UO_2 as temperature and O/U are increased. Figure 43 shows the increase in the diffusion coefficient of oxygen ions in UO_2 as temperature and O/U are increased.

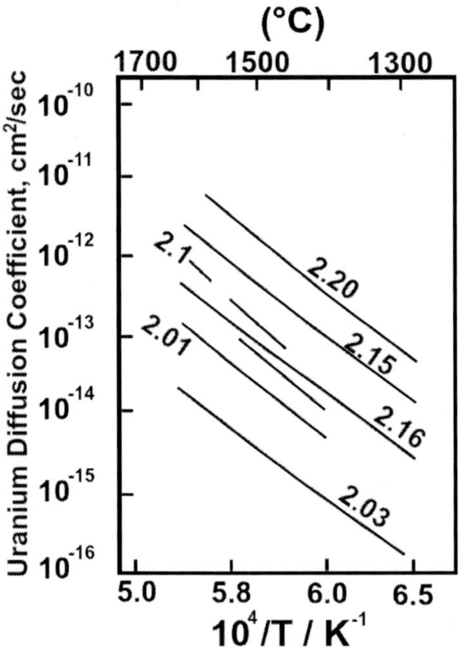

Figure 42. Diffusion coefficient of uranium ion in UO_2 lattice [41].

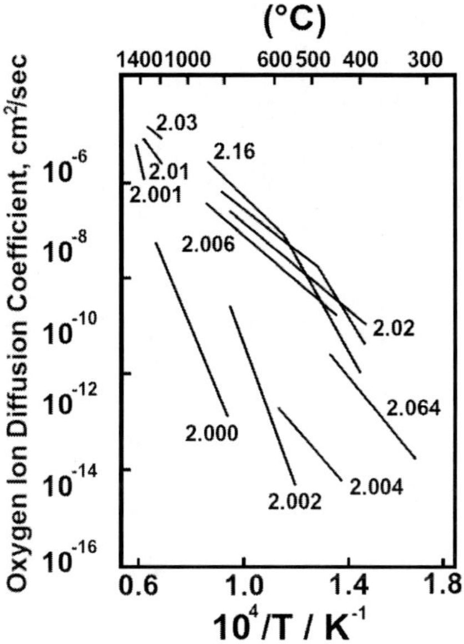

Figure 43. Diffusion coefficient of oxygen ion in UO_2 lattice [41].

The sintering gas hydrogen is sometimes bubbled through water to moisten it. The H_2O in the H_2 gas provides an oxygen potential in the otherwise reducing atmosphere. Figure 44 shows the increase in sintered density of UO_2 versus pH_2/pH_2O.

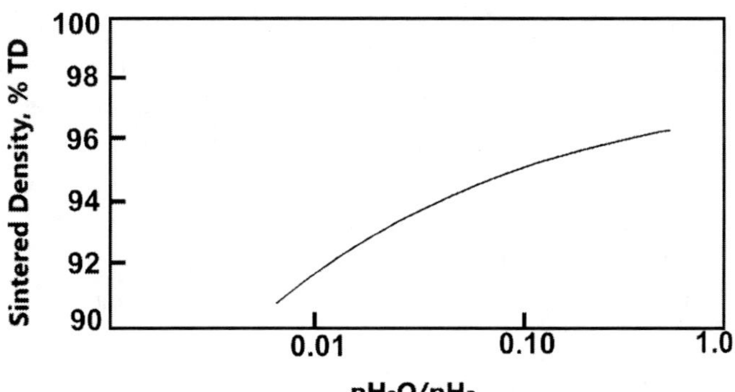

Figure 44. Effect of moisture in hydrogen on density of UO_2 when sintered at 1300°C for 3 hours [32].

8.5. The Effect of Additive/Impurity

Effect of higher valence additive Nb_2O_5

$Nb_2O_5 = 2\ Nb_U^{\bullet} + 4\ O_O + O_i''$

$2\ Nb_2O_5 = 4\ Nb_U^{\bullet} + 10\ O_O + V_U''''$

Effect of lower valence additive Gd_2O_3

$Gd_2O_3 = 2\ Gd_U' + 3\ O_O + V_O^{\bullet\bullet}$

$2\ Gd_2O_3 = 4\ Gd_U' + 6\ O_O + Ui^{\bullet\bullet\bullet\bullet}$

Diffusion rate depends on defect concentrations and their clustering. We have at our disposal several ways of increasing defect concentrations.

1. Increase the temperature.
2. Change the sintering atmosphere.
3. Add a dopant.
4. Subject the material to particle bombardment.

The fourth option of particle bombardment exists in the reactor. The third option of additives carries with it a neutron penalty. The first one may be detrimental to the sintering furnace. Hence changing the atmosphere is an economical option. In case we choose an atmosphere or dopant, the sintering temperature can be brought down. This is used for tiles and ceramics, though not for UO_2 in view of in-reactor thermal stability requirements.

When the oxygen pressure in the sintering furnace atmosphere is on the higher side, we expect a greater concentration of uranium site vacancies, which speed up the movement of uranium ions in the uranium sub lattice. The same effect is realized when UO_2 is doped with a higher valence additive such as Nb_2O_5.

When the oxygen pressure in the sintering furnace atmosphere is on the lower side, we expect a greater concentration of oxygen site vacancies. This means a lower concentration of oxygen interstitials through the Frenkel equilibrium, which in turn means a reduction in the concentration of uranium site vacancies in the uranium sub lattice. The smaller concentration of uranium vacancies leads to a slowdown of the movement of uranium ion in the uranium sub lattice. The same effect is realized when UO_2 is doped with a lower valence additive such as Gd_2O_3.

While using lower valence additives, the deficiency in uranium vacancies is made up by increasing the oxygen pressure. This is achieved by bubbling hydrogen through water or by using a mixture of hydrogen and carbon dioxide or argon with oxygen impurity. Vacuum sintering has also been attempted, which provides some oxygen pressure from the small quantity of residual air after evacuation.

The additive TiO_2 was found to have a remarkable effect on increasing the density of UO_2, even though there is no difference in cation valence. Other mechanisms such as the formation of a liquid phase are believed to promote sintering. Figure 45 shows the sintered density of UO_2 over time. Figure 46 shows the density increase as the sintering temperature is increased.

8.6. Defect Clustering in UO_2

The Willis 2:2:2 defect cluster consists of two vacancies (Vo), two O interstitials displaced in <110> directions (O'), and two other O interstitials displaced in <111> (O"). This is shown in Figure 47.

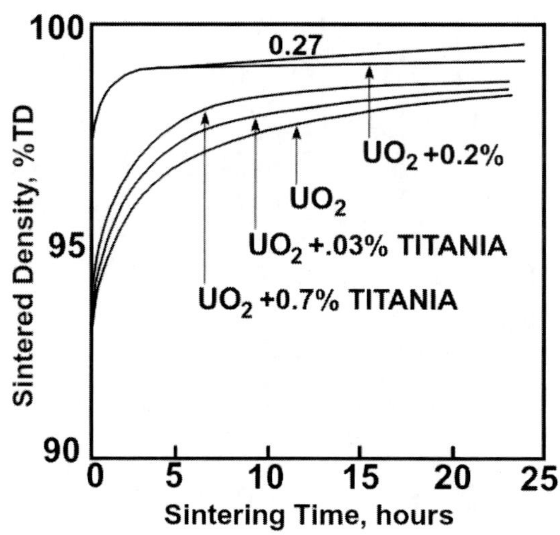

Figure 45. The effect of TiO_2 on densification of UO_2 at 1650°C in hydrogen as a function of time [42].

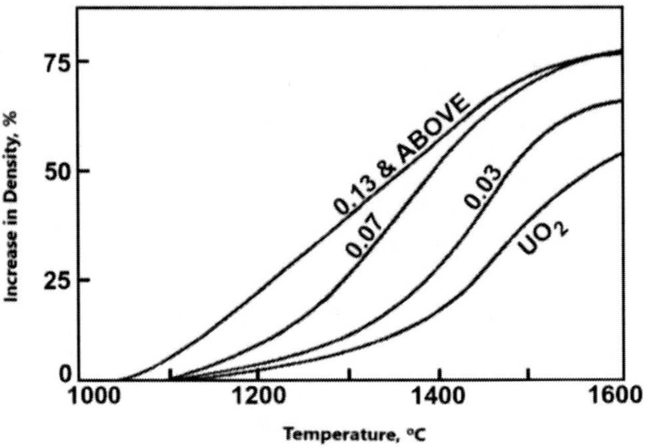

Figure 46. The effect of TiO_2 on densification of UO_2 at 1650°C in hydrogen as a function of temperature [42].

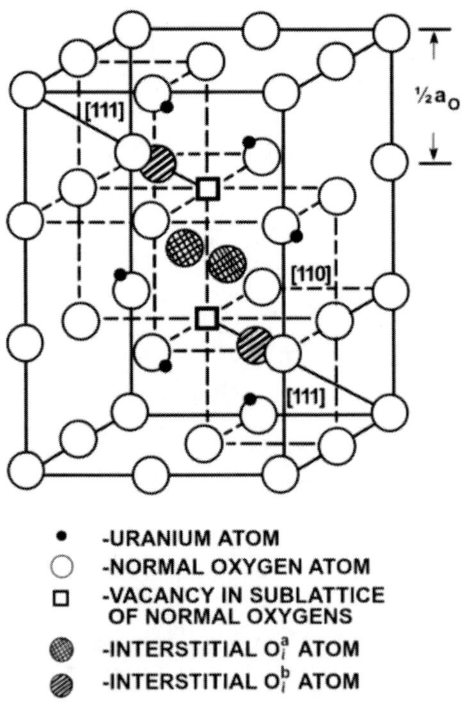

Figure 47. Willis' model of crystal structure of $UO_{2.13}$ [44].

UO_2 is readily oxidized to UO_{2+x} by the incorporation of additional oxygen atoms at interstitial sites of the cubic fluorite structure. For $x < 0.01$, the additional atoms occupy the cubic coordinated sites representing the largest interstices in the fluorite structure. There are four of these in the unit cell. For $x = 0.12$ at 800°C, the additional atoms do not occupy these. They occupy instead two types of interstitial site which are displaced from the cubic coordinate sites by about 1 A in the <110> and <111> directions, respectively, leaving vacant sites in the normal oxygen sub lattice. The interstitial oxygen atoms are not distributed at random throughout the lattice; they cluster together in defect clusters or complexes. Each complex contains <110> and <111> interstitial oxygen atoms, vacant normal oxygen sites and U^{5+} ions to maintain charge balance. There are no displacements of the uranium atoms from their fluorite positions. The most likely configuration for the defect complex in $UO_{2.12}$ is the so called 2:2:2 configuration [43].

9. Defects in Pellets - Agglomerate Effects

The voids in a sintered pellet may originate from different sources. Some of the voids in the green pellet may be larger than normal due to presence of agglomerates and packing difficulties in the pressing die. Inter-agglomerate voids (voids between agglomerates) are larger than intra-agglomerate voids (voids within the agglomerate). Even after shrinkage, these voids remain after sintering. When an additive is involved, for example, Er_2O_3 in UO_2, a larger agglomerate of the additive may leave a large void on dissolution of the additive in the matrix. The dissolution of the additive is complete at temperatures well below that of sintering. Additives to UO_2 may dissolve completely by about 1200°C, whereas densification continues until 1700°C. To reduce the agglomerate size and to improve the packing in the pressing die, the agglomerated powder with no additive (or the powder mixture with agglomerated additive that leaves a void on dissolution) may be milled. In this way, the voids in the green pellet are small and uniform with a good chance of closure in sintering.

9.1. Grain, Particle, Granule, or Agglomerate

Figure 48 illustrates an agglomerate in which particles are joined together. The particle, in turn, consists of several grains or crystals.

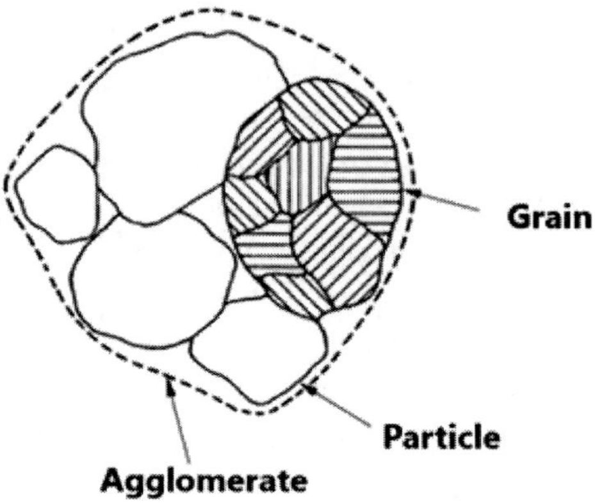

Figure 48. Schematic showing grain, particle, granule, or agglomerate [22].

In Figure 48, the hatched portions represent grains within a particle, the plain shapes represent particles, and the cluster of particles is defined as a granule, if made deliberately, or an agglomerate, if present unintentionally. Granules are regularly shaped as shown in the figure, while agglomerates are irregularly shaped.

9.2. Effect of Agglomerates on Sintered Density

Agglomerates are clusters of particles. Unlike granules which are also clusters of particles, the irregularly shaped agglomerates obstruct powder flow, causing packing problems in pressing and show up as gross defects such as low density or bloating or cracking on sintering. Two categories of powder exhibiting the same BET specific surface area and the same particle size distribution may sinter differently under the same conditions due to the presence of even small quantities of agglomerates.

The increase in sintered density as compaction pressure is shown in Figure 49 for as received and hammer milled UO_2 powders. The as received powder contains agglomerates. Hammer milling reduces the size of the agglomerates in the powder. Note the difference in sintering behavior of compacts from the two powders. When the powder is agglomerate-free, high densities are obtained at relatively lower sintering temperatures. Such figures can be used to determine if the powder is agglomerated.

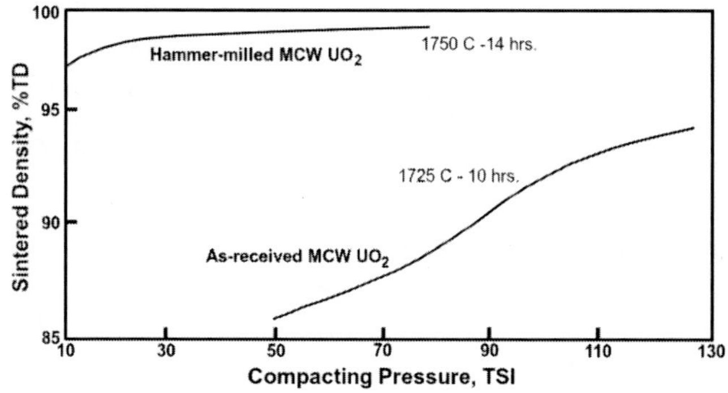

Figure 49. Relationship between compacting pressure and sintered density for as received UO_2 and hammer milled UO_2 powder [45]. When there are no agglomerates, the sintered density is the same over a long pressure range.

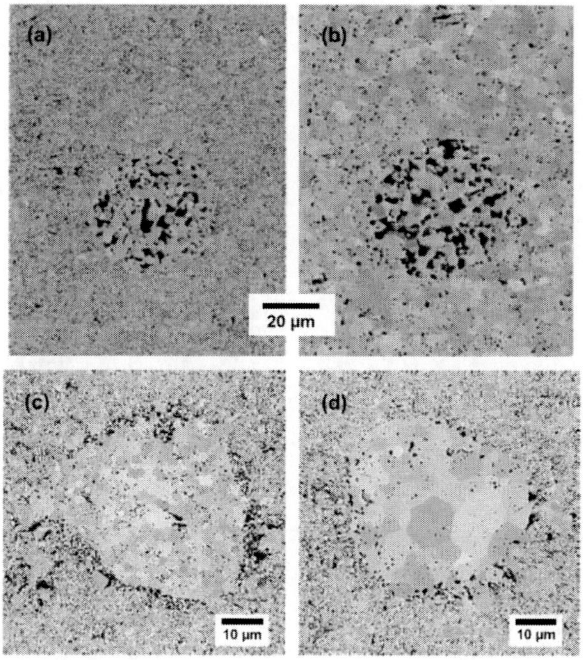

Source: With kind permission from Image Analysis & Stereology, Slovenian Research Agency.

Figure 50. Local microstructure for different defects and sintering conditions: porous aggregate, 1500°C, 1minute. (b) porous aggregate, 1700°C, 1 h. (c) dense aggregate, 1500°C, 1minute. (d) dense aggregate, 1700°C, 1 h. [46].

Girard and co-workers [46], mixed two kinds of agglomerates deliberately into UO_2 powder, pressed and sintered. The first kind is porous aggregate. The second kind is dense aggregate. The microstructures are given in Figures 50(a-b) and 50(c-d), respectively.

9.3. Effects of Physical Inhomogeneity in Green Bodies

It is not only the agglomerates that could cause defects in the sintered pellet. Granules that are made to facilitate flow into the pressing die also can lead to difficulties. The rule of thumb is to restrict the roll pressing or precompaction pressure to 30 to 50% of the final pressing pressure. An additional case can be imagined, where the core of the green pellet happens to have a smaller green density than the periphery. Such a situation can prevail when the powder is too fluffy with large inter-particle friction so that the pressure transmitted to the core is lower than that at the periphery. In such a case, the shrinkage of the core exceeds that of the periphery leading to circumferential cracks.

9.4. Pores Caused by Agglomerates and Agglomerated Additives

Figure 51 shows the density of UO_2 and UO_2-1.5%Gd_2O_3 plotted against UO_2 specific surface area, for the sintering temperatures of 1620 and 1700°C.

Normally, the sintered density is expected to increase as the specific surface area is increased. On the other hand, a dip in the sintered density at higher surface area indicates the presence of agglomerates in the powder. In the case of Figure 51, the agglomerates existed in UO_2 as well as UO_2-Gd_2O_3 powders. The dip in sintered density is evident only when sintered at 1700°C. The Gd_2O_3 agglomerates left larger voids in the sintered pellet that expanded due to trapped gas pressure. The presence of agglomerates was not evident when sintered at 1620°C.

Figure 52 shows a schematic green pellet with several fine pores (say, <5 μm) and one large pore, (say >30 μm) on the left. The green pellet is permeable. On sintering, the small pores have disappeared while the large one remains, as shown in the right-side figure. The sintered pellet is impermeable. Large pores caused by the addition of agglomerated additives can be minimized by milling the powder mixture before pressing [48].

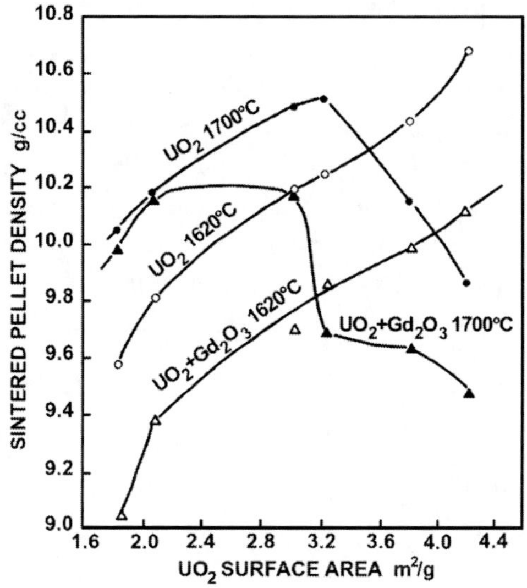

Figure 51. Sintered density vs. BET specific surface area of UO_2 for agglomerated powders [47].

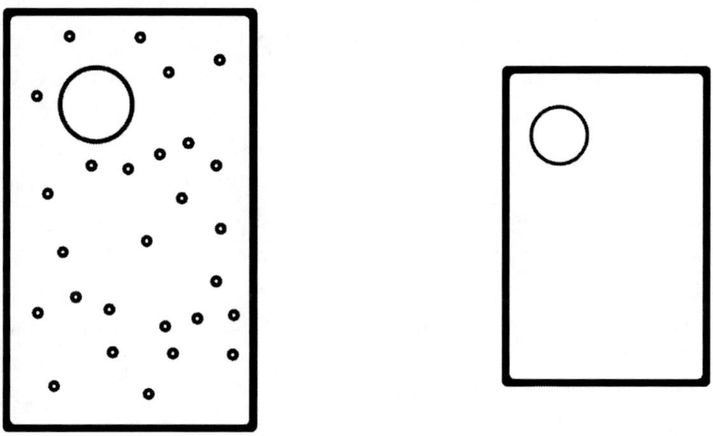

Figure 52. Left: Schematic green pellet with several fine pores (say, < 5μm) and one large pore, (say >30 μm). The green pellet is permeable. Right: Upon sintering, the small pores have disappeared while the large one remains. The sintered pellet is impermeable.

9.5. Caking As a Source of Agglomerates in Powder

The flowability of a fine powder may get reduced further due to caking caused by moisture [49]. The powder particle tends to acquire moisture through the stages of monolayer, multilayer, and finally capillary bridging. The surface tension forces of water bring the particles closer and increase the number of inter-particle contacts. Even though the water is mostly released in thermal treatment, the particles being in close contact in the particle cluster enhance their strength considerably by free sintering. Strong necks (solid bridges) are formed at the contact points.

10. Defects in Pellets – Capping, Cracking, and Unground Surface

Air may be trapped in the pellet during pressing. The compressed air pockets act as separators between regions in the green pellet. The absence of bonding shows up as a capping defect. Clearance between the top punch and die bore permits the escape of air during pressing. Too low a pressing pressure may cause inefficient packing of the powder in the pressing die. A low green density (below a threshold) usually results in a low sintered density. On the other hand, too high a pressing pressure may cause fracture in the green pellet when punch pressure is released or during the ejection of the green pellet from the press die. Stress relief defects such as lamination and end capping may be minimized by proper die design, using a suitable binder, and applying an optimum pressing pressure. A rigid die material such as tungsten carbide, a mirror finish in the die bore, and an exit taper matching the expansion of the green pellet are ideal. Counter-pressure is maintained on the pellet during ejection. Defect 'coring' may be encountered when pellet diameters are large, such as in CANDU fuel, when heating rates are high, or because of impurities from binder burnout.

10.1. Capping/Lamination due to Air Entrapment

These defects are believed to have been caused by air entrapment during pressing. The air between the powder particles also gets compressed along with the powder. When the punch was withdrawn, the air expanded and caused laminations. To permit the escape of air, the diameter of the upper punch tip is kept slightly smaller than the diameter of the lower punch tip. The lower punch tip always stays in the die - down for the fill and up for weight control and through ejection. The lower punch tip stays in the die tightly to keep powder from leaking between the tip and the die wall. The upper punch tip is smaller because this is where the air leaves the die during compression. If the upper punch tips were the same size as the lower punch tips, then the air would be entrapped. The tablet usually caps on the top because that is where the air is pushed out, but if the punch clearance is not enough, capping due to air entrapment can occur elsewhere.

The tendency for the separation of the layers in the compact has been encountered even when there is no air entrapment. This is described in the next sections.

10.2. Stresses in a Compact

Loose powder mass filled in a die undergoes irrecoverable or plastic deformation as the punches compress it from top and bottom. As the applied axial pressure is increased, pressure also gets transmitted radially to the die wall through the powder mass and the forces of reaction from the die wall begin to compress the compact radially. In contrast to metal powder compacts, cold welding is unlikely in ceramic powders which lack ductility. Consolidation takes place by fracture and rearrangement of the fragments. After a certain compactness is achieved, as the applied pressure is increased further, there is some recoverable or elastic deformation in the compact. The compaction die also expands elastically to a small extent depending on die rigidity. Defects in the green compact can originate during the elastic recovery processes of the compact as well as of the die taking place on withdrawal of applied pressure.

Figure 53 shows the stresses in a compact. The vertical compressive stress is maximum at the punch faces and decreases on moving towards the center of the compact. The radial stresses are maximum at the die wall and decrease on moving towards the center due to inter-particle friction. The resultant stresses are maximum at the corners. Stress relaxation is, therefore, likely to be maximum at the corners in opposite directions.

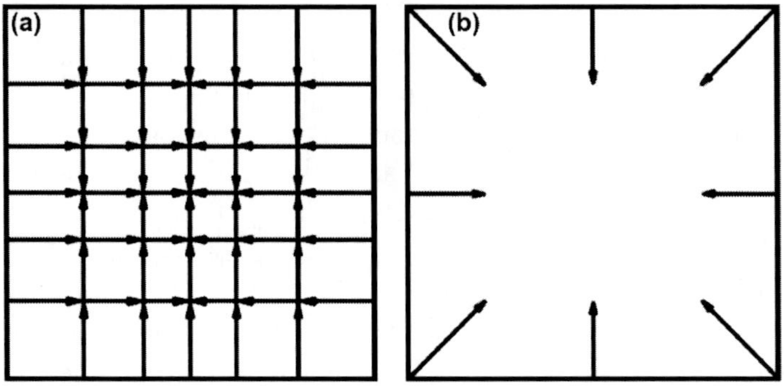

Figure 53. Compaction stresses in a section of a compact in a double acting press (a) axial stresses from applied pressure and radial stresses from die wall reaction (b) resultant stresses [50].

10.3. Lamination due to Insufficient Exit Taper

On removal of the punches after compaction is over and while the compact is still in the die, the compact experiences radial compression by the die wall and tends to expand axially by the Poisson effect. When the axial expansion stress exceeds the green strength, fracture can occur while the compact is still in the die. The radial expansion stress within the compact is balanced by reaction from the die wall as long as the compact is contained in the die, as shown in Figure 54, left. On ejection from the die, the compact tends to expand radially since die wall reaction is no longer available to oppose expansion, as shown in the figure on the right. When the radial expansion stress exceeds the green strength, fracture again occurs as the compact leaves the die. Circumferential laminations appear on the sintered compact on finish grinding.

10.4 End Capping During Decompression

This is found to be due to the inability of the ceramic compact to accommodate stress relief when the upper punch withdraws after compaction. When the compaction pressure is applied, the compact tends to dilate the die sleeve. When the pressure is off, the die sleeve tends to return to its original diameter thereby squeezing the compact. The friction between the die wall and the compact restrains the circumference area from expanding axially while the central portion of the compact is free to expand axially, leading to end capping. This is illustrated in Figure 55.

End capping was found to be less when rigid materials such as tungsten carbide were used in place of cold die steel as the die sleeve material so that the radial squeeze was less during decompression.

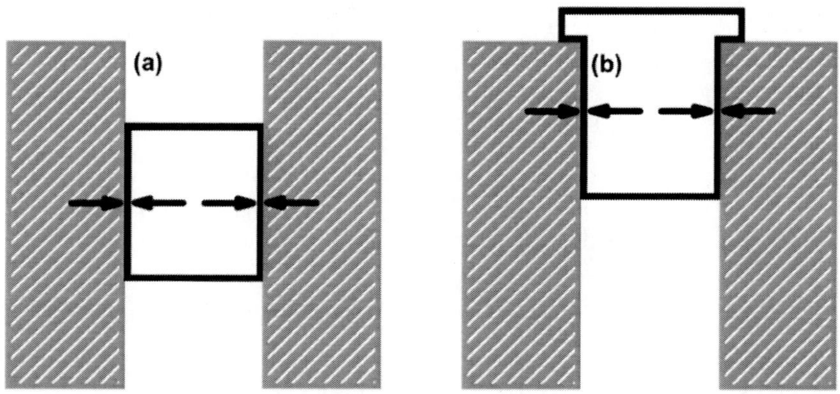

Figure 54. (a) compact within a die on punch withdrawal with radial stress balancing die reaction (b) ejected portion after stress relief and radial expansion [50].

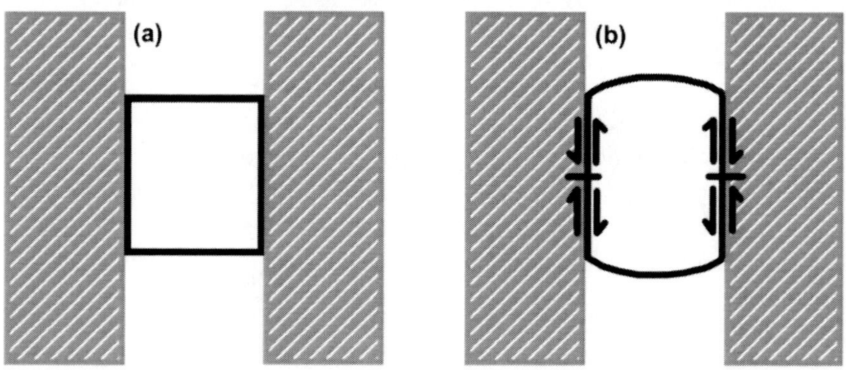

Figure 55. Axial expansion within die of a perfectly elastic compact on punch withdrawal (a) die wall friction zero, (b) die wall friction significant [50].

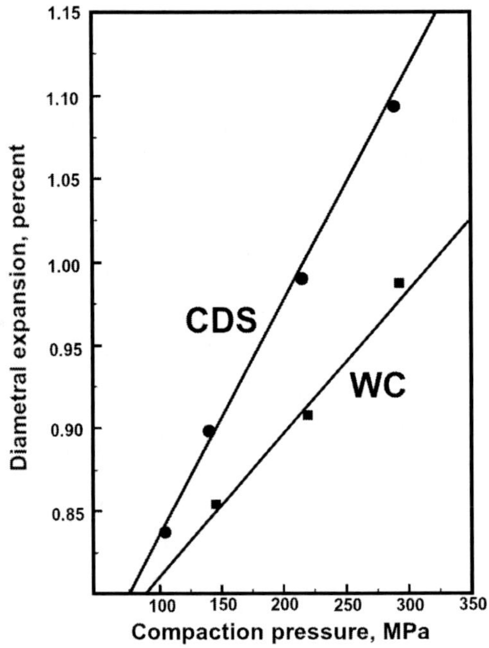

Figure 56. Diametral expansion of a green compact versus compaction pressure in cold die steel (CDS) and tungsten carbide (WC) die sleeves [50].

A simple way to estimate the rigidity of the die sleeve is to measure the diameter of the ejected green pellet. A more rigid die material expands less radially when compaction pressure is applied. Correspondingly, the squeeze on the compact is less when the compaction pressure is withdrawn. Figure 56 gives the experimentally determined diametral expansion of the green compacts made in tungsten carbide and cold die sleeves. It is seen from the figure that for the same compaction pressure, the expansion is larger in the case of cold die steel.

In addition to the use of tungsten carbide die sleeve, the inside of the die bore needs to have a mirror finish to minimize friction enabling uniform axial expansion of the compact at the center as well as at the periphery. In addition, lower compaction pressures are applied, and counter pressure is maintained during ejection of the pressed compact.

It is found that for the same die and pressing parameters, some powder blends yield endcapped pellets while others do not. This could be due to the springiness of the powder arising from the presence of hard agglomerates.

Such powder blends may be reground (hammer milled or attritor milled) to minimize the springiness.

10.5. Unground Patches

The green density may sometimes be less than normal in the central region when top and bottom punches are used. The region will shift when the pressing is from one direction only or when the top and bottom pressing pressures are quite different. Green density variation is minimized in a double acting press using admixed lubricant, pressing to optimum green density, and restricting pellet height to be nearer to the diameter.

When the powder is homogeneous, the low green density central region in the green pellet will shrink more than the higher density edge regions, so that the sintered density is the same in any region. The sintered pellet will then exhibit hourglassing defects, as shown on the left side of Figure 57. When hour-glassed pellets are finish ground, unground patches remain in the central region.

If the mass loss in the pellet on sintering is neglected and if the height of the pellet is taken to be the same as that of the diameter, green and sintered densities are related by

$$d_{sint} = d_{gr} (\rho_{green} / \rho_{sint})^{1/3}$$

where d_{sint} = diameter of the sintered pellet, d_{green} = diameter of the green pellet, ρ_{green} = density of the green pellet, ρ_{sint} = density of the sintered pellet. This expression is useful in estimating the green density that needs to be maintained to prevent unground patches for UO_2 powder blends of known sintered density.

10.6. Other Defects in Uranium Dioxide Pellets

Some of the defects encountered in UO_2 pellets are shown in Figures 58 to 62.

Acceptable quality CANDU pellets are shown in Figure 63.

Figure 57. Hourglass and hour-glassing defect (exaggerated).

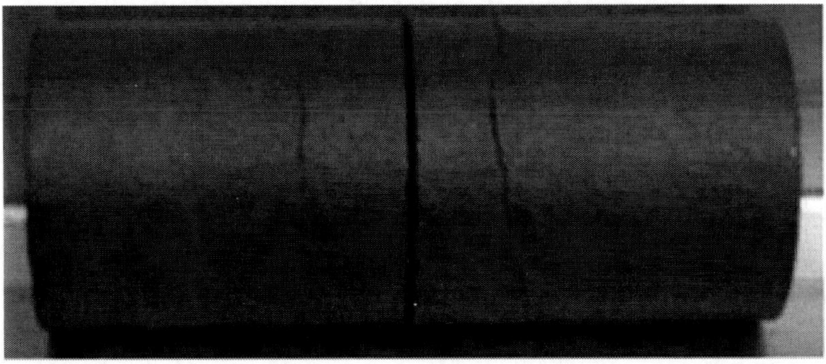

Figure 58. Circumferential cracks in pellets [51].

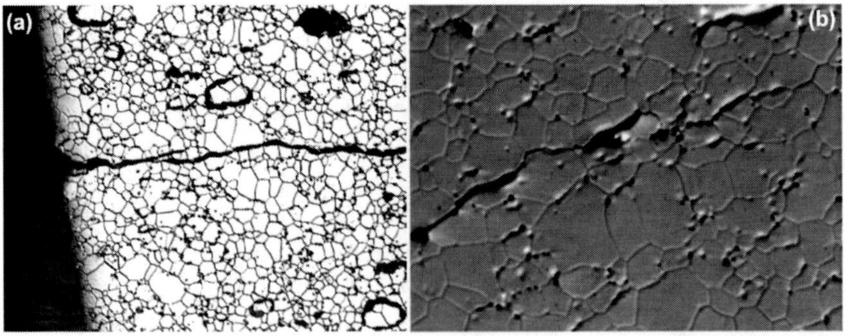

Figure 59. (a) Optical micrograph 100X of the circumferential crack shown in the previous figure. (b) SEM 400 X of the crack [51].

Source: With kind permission from International Atomic Energy Agency.

Figure 60. Coring defect [33].

Source: With kind permission from International Atomic Energy Agency.

Figure 61. End capping [33].

Source: With kind permission from International Atomic Energy Agency.

Figure 62. Fine cracks [33].

Source: With kind permission from International Atomic Energy Agency.

Figure 63. Acceptable quality of CANDU pellets [33].

11. Summary

1. The driving force for sintering is the reduction in surface energy. This surface energy is made available by beginning with very fine particles, with a BET specific surface area of the order of 2 to 4 m^2/g. An aqueous chemical route is commonly used to make fine UO_2 powders, capable of being pressed and sintered to high densities. This aqueous chemical route involves the precipitation of ADU, AUC, or oxalate from the aqueous solution. The bulk density of such powders is 1 to 2 g/cc.
2. The first step in UO_2 powder synthesis, namely, the precipitation of ADU, is controlled to produce fine powders with minimum agglomeration. The reactant concentrations, temperature, pH, and agitation are controlled. Published literature suggests that dilute solutions, low temperatures, low pH, and adequate agitation promote formation of filterable (decantable or centrifugable) precipitates that yield sinterable powders with a low level of agglomeration. Concentrated solutions, high flow rates, high temperatures, and high pH promote the formation of very fine precipitates that agglomerate or coagulate readily. The agglomerates gain strength during thermal treatment, as particle to particle contacts are replaced by solid bridges. They cause flow problems and packing difficulties in the pressing die, and ultimately lead to sporadic low sintered density or even bloating of the sintered body. The original precipitate morphology is retained in the final UO_2 powder. Precipitates from dilute solutions are carefully dried and calcined to achieve a porous structure that readily separates into smaller particles during processing.
3. Some researchers and manufacturers have used two-stage precipitation of ADU from UO_2F_2. The first stage utilized dilute NH_4OH solution (say, 2N) and aimed at producing a filterable (or decantable or centrifugable) precipitate. The second stage used more concentrated NH_4OH solution (say, 10N) and aimed at the total recovery of uranium. Both precipitation vessels used recirculation along with intense agitation by mechanical stirrers to break up agglomerates of ADU.
4. Powders with a narrow particle size distribution (or near monomodal distribution) are expected to result in uniform green microstructures with fine pores that translate to high-density sintered bodies without defects. However, it is not practical to make such powders. Instead,

soft powders with a wide particle size spread are made. The varied sizes are expected to separate into smaller entities in the milling, blending, and pressing processes. In such a case, a pressed compact behaves as if it was made from a narrow PSD powder.

5. The terms 'agglomerates' and 'granules' refer to clusters of particles. However, agglomerates are unwanted and formed by process default. They hinder powder flow. On the other hand, granules are intentionally made to improve flow properties of otherwise nonflowing powder. The bulk density of 1 to 2 g/cc increases on granulation to about 3 g/cc. The granulation may be achieved by roll pressing, or slugging, fragmenting, and sieving or by spray drying, both of which have been used for UO_2 powders.
6. The final pressing operation provides the required shape and inter-particle contact needed for mass transfer during high-temperature sintering. Green densities of the order of 6 g/cc have been sintered to obtain dense UO_2 with a little remnant porosity to accommodate swelling under irradiation.
7. Because UO_2 is nonstoichiometric, the O/U ratio is variable over a wide range, depending on the temperature and oxygen pressure in the environment. Under plant conditions, the powders are hyperstoichiometric before sintering. They become near-stoichiometric on sintering in a reducing atmosphere.
8. Besides agglomerates of UO_2, another source of remnant porosity is additives such as burnable poisons and sintering aids. Examples are Gd_2O_3, Nb_2O_5, TiO_2, Cr_3O_2, and Al_2O_3. If an additive to UO_2 is clumped, it leaves pores on dissolution during sintering. If present as large clumps, additives may behave as unintentional pore formers.
9. The final maximum agglomerate size in the powder is that of the screen size in the hammer mill or attritor mill. By adjusting the milling parameters and reducing the screen size, it is possible to reduce the agglomerate size.
10. Die and press parameters are maintained to minimize stress relief defects such as lamination, end-capping, and hour-glassing. Mirror finished tungsten carbide dies with suitable ejection taper and moderate compaction pressures are favored to avoid pellet losses due to defects.
11. As an alternative to additives, it is possible to promote densification and grain growth in UO_2 by providing a slight oxygen pressure in the sintering furnace atmosphere. This is usually achieved by bubbling

the hydrogen through water. The dew point of the sintering furnace atmosphere may be further increased by heating the water used for bubbling. Mixing hydrogen with carbon dioxide has also been used.

12. Additives may either lower or enhance densification and grain growth. When an additive is found to cause lowering of densification or grain growth (as in the case of grain growth inhibition in erbia-added UO_2), hyperstoichiometry is maintained during sintering to compensate for the adverse effect of the additive. This may be achieved by passing the sintering gas through a bubbler or by mixing CO_2 in the sintering gas. Additives may help densification by maintaining grain boundary pore linkage; for example, MgO. Additives may also help by speeding up the slower diffusing ions; for example, Nb_2O_5.
13. The sintering temperature profile is maintained to permit binder burnout in the preheating zone, additive dissolution in the pre-sintering zone, and completion of densification in the final soaking zone. The net flow of sintering furnace gas is maintained counter to the material movement to sweep away the products of reaction from the binder and excess oxygen from the green pellets. This way, contact of reactants with sintered product is prevented.

Disclaimer

The author has compiled this chapter in his individual capacity as a teacher and learner. It is based solely on published literature. No input of any kind, whatsoever, and no financial or other assistance has been received in any form from any institution or individual. The chapter is primarily meant to provide basic scientific knowledge to the student or the shop floor engineer. The author is not liable for any consequences of interpretations and implementations by the readers.

Acknowledgments

To my former PhD guide Professor P. Ramakrishnan Menon (Emeritus Professor, Indian Institute of Technology, Bombay), my former PhD student, Dr. B. Narasimha Murty, my former colleagues V. Pushpangadan, R. M.

Kartha, D. V. Ratnam, and R. R. K. Raju (Retired, Nuclear Fuel Complex, Hyderabad) for useful discussion.

Editor's Note

Figures in this chapter have undergone graphical alterations/edits to conform to publication standards and are intended for use in this chapter as *illustrations*. The original referenced figures should be used for rigorous scientific examination/comparison.

References

[1] Lee, Y. W., Yang, M. S. Characterization of HWR fuel pellet fabrication using UO_2 powders from different conversion processes. *Journal of Nuclear Materials* 1991, *178*, 217-226.

[2] Rhee, Y. W., Oh, J. S., Kim, D. J., Kim, K. S., Kim, J. H., Yang, J. H., Koo, Y. H. *In-Situ Observation of Sintering Shrinkage of UO_2 Compacts Derived from Different Powder Routes*, Transactions of the Korean Nuclear Society Autumn Meeting Gyeongju, Korea, October 29-30, 2015.

[3] Jarvenin, G. *Precipitation and Crystallization Processes,* Los Alamos National Laboratory, 2009 www.cresp.org/NuclearChemCourse/monographs/09_Jarvinen_FuelCycleSepCrystPrec12-0Sfin_3_2_09.pdf.

[4] Doi, H., Ito, T. Significance of physical state of starting precipitate in growth of uranium dioxide particles. *Journal of Nuclear Materials* 1964, *11*, 94-106.

[5] Manna, S., Roy, S. B., Joshi, J. B. Crystal growth and morphological study of ADU and its calcined product, *BARC Newsletter* 2013, pp. 162-166.

[6] Krishnamoorthy, T. S., Mahadevan, N., Das, M. S. Precipitation of ammonium diuranate – a study, *Proceedings of the International Symposium on uranium technology,* Bombay 13-15 December 1989. Volume 2, Bhabha Atomic Research Centre, 1991.

[7] Janov, J., Alfredson, P. G., Vilkaitis, V. K. The influence of precipitation conditions on the properties of ammonium diuranate and uranium dioxide powders. *Journal of Nuclear Materials* 1972, *4*, 161-174.

[8] Murty, B. N., Palanki, B., Yadav, R. B., Ganguly, C. New approaches in the characterization of uranium dioxide. *Materials Characterization* 2003, *49*, 103-111.

[9] Eloirdi, R., Lin, D. H. M., Mayer, K., Caciuffo, R., Fanhänel, T. Investigation of ammonium diuranate calcination with high temperature X ray diffraction. *Journal of Materials Science* 2014, *49*, 8436-8443.

[10] Runfors, U. The influence of powder characteristics on process and product parameters in UO2 pelletization. *Studsvik,* AE-415, 1971.

[11] Das, P., Choudhury, R. Studies on the sintering behavior of uranium dioxide powder compacts. *Journal of Nuclear Materials* 1988, *158*, 261-266.

[12] Palanki, B., Singh, A. Gupta, U. C., Sinha, K. K. "Agglomerate-free fine ceramic UO_2 powders." In Top Fuel'97, Vol. II, Proceedings of the British Nuclear Energy Society, Thomas Telford, London 1997. Reprinted in *Interceram* 1999, *48*, 98-103.

[13] Pramanik, D., Ravindran, M., Rao, G. V. S. H. "Innovative process technique to optimize quality and microstructure of UO_2 fuel for PHWRs in India," *Advanced Fuel Pellet Materials and Fuel Rod Design for Water Cooled Reactors, (Proceedings of Technical Meeting, 23-26 November 2009, PSI, Villigen, Switzerland)* IAEA-TECDOC-1654, 2010 pp.13-34.

[14] Sandler, N., Reiche, K., Heinämäki, J., Yliruusi, J. Effect of moisture on powder flow properties of theophylline. *Pharmaceutics*, 2010, *2*, 275-290.

[15] Sagel-Ransijn, C. D., Winnubst, A. J. A., Kerkwijk, B., Burggraaf, A. J., Verweij, H. Production of defect-poor nanostructured ceramics of yttria-zirconia. *Journal of the European Ceramic Society* 1997, *17*, 831-841.

[16] Song, K. W., Kim, K. S., Kim, Y. M., Kang, K. W., Jung, Y. H. Reduction of the open porosity of UO_2 pellets through pore structure control. *Journal of Nuclear Materials* 2000, *279*, 253.

[17] Palanki, B., Murty, B. N., Ratnam, D. V., Anuradha, M., Ganguly, C. Light attrition of Uranium dioxide powder, *Ceramics International* 2003, *29*, 99-105.

[18] Palanki, B., Murty, B. N., Yadav, R. B., Anuradha, M., Narayanan, P. S. A., Pramanik, R., Majumdar, S. Admixed binders and lubricants in ceramic powder pressing – a study of zinc behenate additive. *Indian Journal of Engineering and Material Sciences* 1998, *5*, 136-139.

[19] Yanai, K., Hirai, M., Ishikawa, T., Ishizaki, J., Saitoh, H. An analysis of density distribution in UO_2 green pellet by finite element method. *Journal of Nuclear Materials* 1998, *257*, 318-330.

[20] Porneuf, A., Hausner, R. "Meeting on characterization of uranium dioxide." *ORNL,* December 12-13, 1961, TID-7637, PP.167-189.

[21] Jeon, S. C., Lee, J. W., Kang, S. J., Lee, J. H., Lee, J. W., Park, G. I., Kim, I. T. Temperature dependences of the reduction kinetics and densification behavior of U_3O_8 pellets in Argon atmosphere. *Ceramics International* 2015, *41*, 657-662.

[22] Burke, J. E., Rosolowski, J. H. "Section 10 - Sintering" in: *Treatise on Solid State Chemistry, Vol. 4, Reactivity of Solids*, N. B. Hannay (Ed.), Plenum Press, 1976.

[23] German, R. M. Sintering trajectories: Description on how density, surface area and grain size change. *Journal of Metals* 2016, *68*, 878-884.

[24] Irvine, J. T. S. *"Processing of Materials: Lecture JI4 – Sintering of Ceramics II,"* CH5716, University of St. Andrews, UK, 2015.

[25] De Jonghe, L. C., Rahaman, M. N. "4.1 Sintering of Ceramics" In: *Handbook of Advanced Ceramics*, S. Somiya, F. Aldinger, R. M. Spriggs, K. Uchino, K. Koumoto, M. Kaneno (Eds.), Elsevier Inc. 2003, p. 247.

[26] Weidinger, H., Assmann, H., Holzer, R., Raab, K. "Fuel rod fabrication and engineered quality assurance under the requirements of KWU design," *IAEA-SM-233/17* in: Fabrication of water reactor fuel elements, 1979, pp. 151-176.

[27] Heal, T. J., Littlechild, J. E., Watson, R. H. "Development of stable density UO_2 fuel," *Nuclear Fuel Performance,* British Nuclear Energy Society, 1973, pp. 52.1-52.3

[28] Kang, S. J. L. *Sintering: Densification, Grain Growth and Microstructure*, Elsevier 2005.

[29] Barringer, E., Jubb, N., Fegley, B., Pober, R. L., Bowen, H. K. "Processing monosized powders" in: *Ultrastructure processing of ceramics, glasses and composites*, L. L. Hench and D. R. Ulrich (Eds.), Wiley Interscience 1984.

[30] Brook, R. J. Pore-grain boundary interactions and growth, *J. Am. Chem. Soc.* 1969, *52*, 56-57.

[31] Palanki, B., Murty, B. N., Anuradha, M., Yadav, R. B., Jayaraj, R. N. Incidence of high nitrogen in sintered uranium dioxide. *Nuclear Technology* 2005, *150*, 189-195.

[32] Stuart, W. I., Adams, R. B. Effect of sintering atmosphere on the density of uranium dioxide pellets. *Journal of Nuclear Materials* 1975, *58*, 201-201.

[33] Pramanik, D., Ravindran, M., Rao, G. V. S. H. Nuclear Fuel Complex, *Presentation,* 2009.

[34] Alexander, B. H., Balluffi, R. W. The mechanism of sintering of copper. *Acta Metallurgica* 1957, *5*, 666-677.

[35] Palanki, B. Some factors affecting densification and grain growth in the sintering of uranium dioxide – a brief review. *Journal of Nuclear Materials* 2021, *550*, 152918.

[36] Kingery, W. D., Francois, B. "The sintering of crystalline oxides, I. Interactions between grain boundaries and pores," in: *Sintering and Related Phenomena*, G. C. Kuczynski, N. Hooten & C. Gibson (eds.), Gordon & Breach Science Publishers, New York, 1967, pp. 471-96.

[37] Hausner, H. "Grain growth of UO_2 Part –II," Joint US Euratom Research and Development Program, GEAP 4689, Oct 1965.

[38] MacEwan, J. R. Grain growth in sintered uranium dioxide: 1, Equi-axed grain growth. *Journal of American Ceramic Society* 1962, *45*, 37-41.

[39] Kofstad, P. *Non-stoichiometry, Diffusion and Electrical Conductivity in binary metal oxides*. John Wiley, 1972.

[40] Desgranges, L., Gramond, M., Petot, C., Petot-Evans, G., Ruello, P., Saadi, B. Characterization of uranium vacancies in hyper stoichiometric uranium dioxide. *Journal of the European Ceramic Society* 2005, *25*, 2683–2686.

[41] Keiji, N., Naoki, K. "High temperature chemistry of ceramic nuclear fuels with emphasis on nonstoichiometry," in: *Advances in Nuclear Science and Technology*, Vol. 9, Edited by E. J. Henley and J. Lewins, Academic Press, New York, 1976, pp. 99-180.

[42] Ainscough, J. B., Rigby, F., Osborn, S. C. The effects of titania on grain growth and densification of sintered UO_2 *Journal of Nuclear Materials* 1974, *52*, 191-203.

[43] Willis, B. T. M. The defect structure of hyper stoichiometric uranium dioxide. *Acta Crystallographica* 1978, *A34*, 88-90.

[44] Willis, B. T. M. Positions of the oxygen atoms in $UO_{2.13}$ *Nature* 1963, *197*, 755.

[45] Schaner, B. E. "4.2.1 Cold Pressing," in: *Uranium Dioxide, Properties and Nuclear Applications*, J. Belle (Ed.), United States Atomic Energy Commission, 1961. Page 129.

[46] Girard, E., Chaix, J. M., Valdivieso, F., Goeuriot, P., Lechelle, J. Local study of defects during sintering of UO_2: image processing and quantitative analysis tools. *Image Analysis and Stereology* 2008, *27*, 79-85.

[47] Palanki, B., Kulkarni, A. P., Krishnan, T. S., Balaramamoorthy, K., Ramamohan, T. R., Ramakrishnan, P. "Retarded densification and de-sintering in ceramic bodies" In *Advanced Ceramics*. P. Ramakrishnan (Ed.), Oxford & IBH Publishing Co., New Delhi, 1992, pp.67-76.

[48] Palanki, B. Fabrication of UO_2-Gd_2O_3 pellets, *Journal of Materials Science and Chemical Engineering*, 2016, *4*, 8-21.

[49] Paternina, J. Preventing moisture caking, the unwanted agglomeration. *PBE* March 2014 www.powderbulk.com .

[50] Palanki, B., Chakraborthy, K. P., Singh, A. End capping and other defects in pressed ceramic compacts, *Indian Journal of Engineering and Material Sciences*, 1996, *3*, 196-200.

[51] Oliveira, D. S., Paneto, L. F. P. C., de Souza, P. O. *Study of transverse crack formation on surface area of UO_2 pellet circumference*, International Nuclear Atlantic Conference – INAC 2013, Recife, PE, Brazil, November 24-29, 2013.

Further Reading

Abdeirazek, D., Zidan, W. I. "Apparent Chemical Reactions Kinetics of the Ammonium Diuranate Precipitation," *Sixth Arab Conference on the Peaceful uses or Atomic Energy.* Cairo, Egypt, 4 - 8 Dec. 2002.

Advances in Uranium refining and conversion, *IAEA TECDOC* 420, 1987.

Alfredson, P. G. "Pilot plant development of processes for the production of nuclear grade uranium oxide," *AAEC*/ E245 Nov. 1972.

Alfredson, P. G., Janov, J. "Investigation of batch tray calcination – reduction of ammonium diuranate to uranium dioxide," *AAAC* TM 599, 1971.

Amelinckx, S., Bressers, J., Nevelsteen, K., Smets, E., Van Lierde, W. Growth of UO_2 single crystals. *Euratom, Eur* 345.e, 1963.

Araujo, J. A. de, Study of continuous precipitation of ADU for the implantation in the pilot installation, Report No. IEA-339 (in Portuguese), *Institute of Atomic Energy, Brazil* 1974.

Arthur J. Miller, A. J. *Precipitation of ammonium diuranate,* US Patent 2466118, Application April 5, 1949.

Assmann, H., Dorr, W., Peehs, M. Control of UO_2 microstructure by oxidative sintering. *Journal of Nuclear Materials* 1986, *140*, 1-6.

Assmann, H., Peehs, M., Roepenack, H. Survey of binary oxide fuel manufacturing and quality control. *Journal of Nuclear Materials* 1988, *153*, 115-126.

Bahraminasab, M., Farahmand, F. State of the art review on design and manufacture of hybrid biomedical materials: Hip and knee prostheses. *Proc. Inst. Mech. Eng. H* 2017 Sep *231*, 785-813.

Barry, C. C., Grant, N. M. *CERAMIC MATERIALS – Science and Engineering*, Springer 2007, p.91, pp. 430-432.

Barsoum, M. W. *Fundamentals of Ceramics*, Institute of Physics, *Series in Materials Science and Engineering*, 2002 p. 316.

Bjork, R., Tikare, V., Frandsen, H. L., Pryds, N. The sintering behavior of close packed spheres. *Scripta Materialia* 2012, *67*, 81–84.

Blum, A. *The manufacture of uranium dioxide fuel in pellet form,* INIS-MF-66/12, 1966.

Bourgeois, L. Dehaudt, Ph., Lemaignan, C., Hammou, A. Factors governing microstructure development of Cr_2O_3 doped UO_2 during sintering. *Journal of Nuclear Materials* 2001, *297*, 313-326.

Bromley, A. P., Logsdon, R., Roberts, V. A. "Changes in UO_2 powder properties during processing via BNFL's binder less route," *IAEA TECDOC-941*, 1997, pp. 99-106.

Bryson, A. W. Electrolytic precipitation of ammonium diuranate, *Journal of the South African Institute of Mining and Metallurgy*, 1975, *9*, 13-15.

Cardinaels, T., Hertog, J., Vos, B., de Tollenaere, L., Delafoy, C., Verwerft, M. Dopant solubility and lattice contraction in gadolinia and gadolinia–chromia doped UO_2 fuels. *Journal of Nuclear Materials* 2012, *424*, 289–300.

Carrea, A. J. Sintering of uranium dioxide in an atmosphere of controlled hydrogen content. *Journal of Nuclear Materials* 1963, *2*, 275-277.

Castell, M. R. Wulff shape of microscopic voids in UO_2 crystals. *Physical Review B* 2003, *68*, 235411.

Chou, K. S., Lin, D. Y. Precipitation studies of Ammonium Uranyl Carbonate from UO_2F_2 solutions. *Journal of Nuclear Materials* 1989, *165*, 171-178.

Coble, R. L. Sintering crystalline solids. I. Intermediate and final state diffusion models. *Journal of Applied Physics*, 1961, *32*, 787-792.

Cooper, M. D. W., Stanek, C. R., Andersson, D. A. The role of dopant charge state on defect chemistry and grain growth of doped UO_2. *Acta Materialia* 2018, *150*, 403-413.

Cordfunke, E. H. P., Van Der Giessen, A. A. Particle properties and sintering behaviour of uranium dioxide. *Journal of Nuclear Materials* 1967, *24*, 141-149.

Crowe, G. A. Precipitation of dense ammonium diuranate, *Spencer Chem Co. US Patent* 3,000,698, September 19, 1961.

Das, P., Choudhury, R. Sintering studies of urania powder compacts – analysis of a flow model. *Journal of Nuclear Materials* 1990, *170*, 286-289.

de Assis, G., Ferraz, W. B., Tambourgi, E. B. "Aspects on sintering and grain growth in pure and Al_2O_3-SiO_2 doped UO_2 pellets." *International Nuclear Atlantic Conference – INAC* 2009, Rio de Janeiro, Brazil, Sept 27- Oct 2, 2009.

de Freitas, C. T. "Uranium dioxide sintering kinetics and mechanisms under controlled oxygen potentials," *Doctoral Thesis*, Instituto De Energia Atomica, Sao Paulo, Brazil, June 1980.

De Hollander, W. R. *Method of preparing uranium diuranate*, GE Patent US 3,394,997, July 30, 1968.

Desson, M., Michel, P., Rerolle, C. "Research, Development and Industrial Experience in Simo in the Precipitation of Ammonium Diuranate" in: *Crystallization and Precipitation, Proceedings of the International Symposium*, Saskatoon,

Saskatchewan, Canada, 5-7 October 1987, G. L. Strathdee, I I Klein, Lamelis (Eds.), Pergamon Press, Oxford. pp. 159-165.

Dimokrati, A. "Phase field simulation of normal grain growth." *International Congress of Automotive and Transport Engineering, CONAT 2016*, Transilvania University of Brasov, 2016. ISSN 2069-0401.

Divins, L. A., Runion, H. L. *Continuous process for the production of powdered uranium dioxide from uranyl nitrate*, GE Patent US 4,656,015 April 7, 1987.

Dooies, B. J. "Enhancement of uranium dioxide thermal and mechanical properties by oxide dopants," *MS Dissertation*, University of Florida, 2008.

Ferreira, R. A. N., Lopes, J. A. M. "Implementation of a fuel pellets density and open porosity measurement system by the xylol penetration immersion method at CDTN," 2007 *International Nuclear Atlantic Conference - INAC 2007*, Santos, SP, Brazil, Sep 30 - Oct 5, 2007.

Francois, B., Kingery, W. D. "The sintering of crystalline oxides, II. Densification and Microstructure Development in UO_2," in: *Sintering and Related Phenomena*, G. C. Kuczynski, N. Hooten & C. Gibson (eds.), Gordon & Breach Science Publishers, New York, 1967, pp. 499-525.

Frost, B. R. T. *Nuclear Fuel Elements, Design, Fabrication and Performance*, Pergamon Press 1982.

Gao, J., Yang, X., Li, R., Wang, Y., Zhong, F. Low-temperature sintering mechanism on uranium dioxide. *Journal of Materials Science* 2007, *42*, 5936-5940.

GE Patent US 4,234,550 November 18, 1980.

Ge, L., Subhash, G., Baney, R. H., Tulenko, J. S., McKenna, E. Densification of uranium dioxide fuel pellets prepared by spark plasma sintering (SPS). *Journal of Nuclear Materials* 2013, *435*, 1-9.

Geetha, M. S., Nagabhushana, H., Shivananjaiah, H. N. Green mediated synthesis and characterization of ZnO nanoparticles using Euphorbia Jatropa latex as reducing agent. *Journal of Science: Advanced Materials and Devices* 2016, *1*, 301–310.

German, R. M. A measure of the number of particles in agglomerates. *International Journal of Powder Metallurgy* 1996, *32*, 365-373.

Glass, S. J., Ewsuk, K. G. Ceramic powder compaction. *MRS Bulletin* 1997, *22*, 24-28.

Glodeanu, F., Spinzi, M., Balan, V. Correlation between UO_2 powder and pellet quality in PHWR fuel manufacturing. *Journal of Nuclear Materials* 1988, *153*, 156-159.

Greenquist, I., Tonks, M. R., Zhang, Y. Review of sintering and densification in nuclear fuels: Physical mechanisms, experimental results, and computational models. *Journal of Nuclear Materials* 2018, *507*, 381-395.

Harvey, M. R., Teter, A. R., Leggett, R. L. Fabrication of oxide nuclear fuel pellets, The Dow Chemical Company, *REP – 1255*, March 21, 1969.

Hasegawa, S. I., Takano, E., Sekine, M. Process for manufacturing uranium dioxide powder, *Mitsubishi Patent US 4*,505,882 March 19, 1985.

Hingant, N., Clavier, N., Dacheux, N., Hubert, S. S., Barre, N., Podor, R., Aranda, L. Preparation of Morphology Controlled $Th_{1-x}U_xO_2$ Sintered Pellets from low-Temperature Precursors. *Powder Technology* 2011, *208*, 454-460.

Hoyt, R. C. "Precipitation kinetics of a continuous precipitator, with application to the precipitation of ammonium poly uranate," *Ph. D Thesis*, Iowa State University, 1978.

Huang, C. T., Kuo, M. S., Tyen, J. G., Shieh, M. C. Process for converting uranyl compounds to UO_2 via ADU, *Taiwan Patent US 5*,139,709 November 21, 1989.

Hung, N. T., Thuan, L. B., Tung, N. V., Thuy, N. T., Lee, J. Y., Jyothi, R. K. The UO_2 ex-ADU powder preparation and pellet sintering for optimum efficiency: experimental and modeling studies," *Journal of Nuclear Materials* 2017, *496*, 177-181.

Ikegami, T., Moriyoshi, Y. Intermediate Stage Sintering of a Homogeneously Packed Compact. *Journal of the American Ceramic Society* 1984, *67*, 174-178.

Imakuma, K., Riella, H. G., Lordello, A. R., Iyer, S. S., Sabato, S. F., Grigoletto, T., Vea, O., Salvador, V. L., Kakazu, M. H., Pires, M. A. F., De Lima, N. B. Fuel characterization techniques: current methods practiced at the Instituto De Pesquisas Energeticas E Nucleares, Brazil, *Journal of Nuclear Materials* 1988, *153*, 76-81.

Jeon, S. C., Lee, J. W., Lee, J. H., Kang, S. J., Lee, K. Y., Cho, Y. Z., Ahn, D. H., Song, K. C. Fabrication of UO_2 Porous Pellets on a Scale of 30 kg-U/Batch at the PRIDE Facility. *Advances in Materials Science and Engineering* 2015, 376173.

Joshi, S. S. *Crystal Habit Modification Using Habit Modifiers, Modern Aspects of Bulk Crystal and Thin Film Preparation*, Nikolai Kolesnikov (Ed), 2012, InTech

Joung, C. Y., Lee, S. C., Kim, S. H., Kim, H. S., Sohn, D. S. Fabrication method for UO_2 pellets with large grains or a single grain by sintering in air. *Journal of Nuclear Materials* 2008, *375*, 209–212.

Jovani-Abrill, R. "Synthesis and characterization of nano crystalline UO_2 ceramics," *Doctoral Thesis,* Santiago de Compostela, Spain 2014.

Kim, E. H., Park, J. J., Park, J. H., Chang, I. S., Choi, C. S., Kim, S. D. Thermal decomposition kinetics of ammonium uranyl carbonate. *Journal of Nuclear Materials* 1994, *209*, 294-300.

Kingery, W. D. Implications of sintering theories with regard to process controls, Transactions of the VII International Ceramic Congress, London, Institute of Ceramics, Stoke-on-Trent, 1960. pp. 461-71.

Kingery, W. D., Francois, B. "The sintering of crystalline oxides, I. Interactions between grain boundaries and pores," in: *Sintering and Related Phenomena*, G. C. Kuczynski, N. Hooten & C. Gibson (eds.), Gordon & Breach Science Publishers, New York, 1967, pp. 471-96.

Kingery, W. D., Francois, B. J. Grain growth in porous ceramics. *Journal of the American Ceramic Society* 1965, *48*, 546-547.

Knudsen, I. E., Hootman, H. E., Levitz, N. M. A fluid bed process for the direct conversion of uranium hexafluoride to uranium dioxide, ANL-6606, Technology - Feed Materials TID-4500, 19th Ed.) *AEC Research and Development Report*, Feb 1963.

Kofstad, P., Norby, T. Defects and transport in crystalline solids, Advanced level course, University of Oslo, 2004.

Kong, L. B., Huang, Y., Que, W., Zhang, T., Li, S., Zhang, J., Dong, Z., Tang, D. "Sintering and Densification (I) – Conventional Sintering Technologies," in: *Transparent Ceramics*, Springer, 2015, 291-394.

Kuzmin, R. S., Baranov, V. G., Burlakova, M. A., Mikhalchik, V. V., Tenishev, A. V. Rate control sintering of uranium dioxide, *IOP Conf. Series: Materials Science and Engineering* 2016, *130*, 012021.

Lambert, J. D. B., Strain, R. "Oxide Fuels." In Materials Science and Technology, (eds R. W. Cahn, P. Haasen and E. J. Kramer). 2006.

Lay, K. W., Carter, R. E. Role of the O/U on the sintering of UO_2, *Journal of Nuclear Materials* 1969, *30*, 74-87.

Leckie, R. M., Luther, E. P. "*Report on Controlling the Microstructure of UO_2 Pellets*," *LA-UR*-12-24919, 2012.

Lee, M. C., Wu, C. J. Conversion of UF_6, to UO_2: A quasi-optimization of the ammonium uranyl carbonate process. *Journal of Nuclear Materials* 1991, *185*, 190-201.

Lee, Y. W., Yang, M. S. Characterization of HWR fuel pellet fabrication using UO_2 powders from different conversion processes. *Journal of Nuclear Materials* 1991, *178*, 217-226.

Lee. S. J., Kim, K. T. "The Influence of the Sintering Atmosphere on Properties of Refractories for UO_2 Sintering Furnace and Microstructure of UO_2 Pellets." *The 2nd Seminar on the New Fuel Technology Toward the 21st Century*, November 25 - 26, 1997, KAERI, Taejon, Korea.

Luther, E., Necker, C., Mihaila, B., Papin, P., Guirdy, D. "Microstructural characterization of uranium oxide," *LA-UR-11-00241, 10-08398, Nuclear Fuels and Materials, American Nuclear Society*, June 26-30, 2011.

Maier, G., Assmann, H., Dorr, W. Resinter testing in relation to in-pile densification. *Journal of Nuclear Materials* 1988, *153*, 213-220.

Mann, M., Hunt, E., Young, C., Kimani, M., Turner, D., Varga, S., Petrosky, J. Hydrothermal growth and characterization of UO_2 single crystals for neutron radiation detection. *Proceedings Volume 9968, Hard X-Ray, Gamma-Ray, and Neutron Detector Physics XVIII; 99680R;* SPIE Optical Engineering + Applications, 2016, San Diego, California, United States.

Manna, S., Basak, C., Thakkar, U. R., Tahkur, S. A., Roy, S. B., Joshi, J. B. Study on effect of process parameters and mixing on morphology of ammonium diuranate, *Journal of Radioanalytical and Nuclear Chemistry* 2016, *310*, 287-299.

Manna, S., Hareendran, K. N., Banerjee, J., Kutty, T. R. G., Roy, S. B., Joshi, J. B. Study of Microstructure of Ammonium Di Uranate and Uranium Oxide Produced from Differently Sourced Ore, *Proceedings of GLOBAL 2011*, Makuhari, Japan Dec 11-16, 2011. Paper No. 357682.

Manna, S., Karthik, P., Mukherjee, A., Banerjee, J., Roy, S. B., Joshi, J. B. Study of calcinations of ammonium diuranate at different temperatures. *Journal of Nuclear Materials* 2012, *426*, 229-232.

Manna, S., Kumar, R., Satpati, S. K., Roy, S. B., Joshi, J. B. Study of the changes in composition of ammonium diuranate with progress of precipitation and study of the properties of ammonium diuranate and its subsequent products produced from both uranyl nitrate and uranyl fluoride solutions. *Nuclear Engineering and Technology* 2017, *49*, 541-548.

Manna, S., Roy, S. B., Joshi, J. B. Study of crystallization and morphology of ammonium diuranate and uranium oxide. *Journal of Nuclear Materials* 2012, *424*, 94-100.

Marajofsky, A., Perez, L., Celora, J. On the dependence of characteristics of powders on the AUC. *Journal of Nuclear Materials* 1991, *178*, 143-151.

Martinez, J., Clavier, N., Ducasse, T., Mesbah, A., Audubert, F., Corso, B., Vigier, N., Dacheux, N. From uranium (IV) oxalate to sintered UO_2: Consequences of the powders' thermal history on the microstructure. *Journal of the European Ceramic Society* 2015, *35*, 435-4546.

Mayo, M. J. "Nanocrystalline Ceramics for Structural Applications: Processing and Properties." In: *Nanostructured Materials*. 1998 vol 50 Chow, G. M., Noskova, N. I. (Eds.) Springer, Dordrecht.

McCoy, J. J. *Apparatus for producing UO_2 powder GE Patent US* 3,734,694, May 22, 1973.

Mellah, A., Chegrouche, S., Barkat, M. The precipitation of ammonium uranyl carbonate (AUC): Thermodynamic and kinetic investigations. *Hydrometallurgy* 2007, *85*, 163-171.

Mihaila, B., Luther, E. P. *"MSC – DFEM Model for the Sintering of Ceramic Nuclear Fuels,"* LA-UR-12-25082, 2012.

Moore, E., Guéneau, C., Crocombette, J. P. Diffusion model of the non-stoichiometric uranium dioxide, *Journal of Solid State Chemistry* 2013, *203*, 145–153.

Murty, B. N., Palanki, B., Yadav, R. B., Ganguly, C. Influence of temperature of precipitation on agglomeration and other powder characteristics of ammonium diuranate. *Powder Technology* 2001, *115*, 167-183.

Nishijima, T., Kawada, T., Kitagawa, H. An improved process for fabricating UO_2 pellets, *Journal of Nuclear Materials* 1965, *16*, 184-189.

Oberacker, R. "Powder compaction by dry pressing," in: *Ceramics Science and Technology, Volume 3: Synthesis and Processing*, Ralf Riedel and I-Wei Chen (Eds.), Wiley-VCH Verlag GmbH, 2012, p.15.

Oguma, M. Microstructure Effects on Fracture Strength of UO_2 Fuel Pellets. *Journal of Nuclear Science and Technology* 1982, *19*, 1005-1014.

Oh, J. S., Kim, D. J., Yang, J. H., Kim, K. S., Rhee, Y. W., Koo, Y. H. Recycling process of U_3O_8 powder in MnO-Al_2O_3 doped large grain UO_2 pellets. *Nuclear Engineering and Technology* 2014, *46*, 117-124.

Oolman, T. "Characterization of ammonium polyuranate powders from a continuous precipitator," *MS Dissertation,* Iowa State University, 1979.

Paik, S., Biswas, S., Bhattacharya, S., Roy, S. B. Effect of ammonium nitrate on precipitation of ammonium diuranate and its characteristics. *Journal of Nuclear Materials* 2013, *440*, 34-38.

Palanki, B., Murty, B. N., Sahoo, P. K., Gopalakrishna, T. Green strength of zirconium sponge and uranium dioxide powder compacts. *Materials Characterization* 2008, *59*, 871-876.

Palanki, B., Asnani, C. K., Kartha, R. M., Ramachandran, K., Babu, K. S., Ravichandran, V., Murty, B. N., Ganguly, C. Uranium dioxide powder preparation, pressing, and sintering for optimum yield. *Nuclear Technology* 1999, *127*, 375-381.

Palanki, B., Murthy, B. N., Anuradha, M. A new process-based agglomeration parameter to characterize ceramic powders. *Journal of Nuclear Materials* 2009, *384*, 190–193.

Palanki, B., Murty, B. N., Chakraborthy, K. P., Jayaraj, R. N., Ganguly, C. Coarsening densification transition temperature in sintering of uranium dioxide. *Journal of Nuclear Materials* 2001, *297*, 35-42.

Pan, Y. M., Ma, C. B., Hsu, N. N. The conversion of UO_2 via ammonium uranyl carbonate: study of precipitation, chemical variation and powder properties. *Journal of Nuclear Materials* 1981, *99*, 135-147.

Patton, F. S., Gogan, J. M., Griffith, W. L. *Enriched Uranium Processing*, Oxford 1963, pp.47-56.

Production of yellow cake and uranium fluorides, *IAEA* 1980.

Rahaman, M. N. *Ceramic Processing and Sintering*, 2nd Edition, Marcel Dekker, New York 1995.

Rahaman, M. N. *Ceramic Processing and Sintering*, CRC Press: Boca Raton, 2003.

Rajagopal, S., Ansari, T. P. S., Iyer, C. S. P. Effect of ammonium nitrate on precipitation of ammonium diuranate and its characteristics. *Journal of Nuclear Materials* 1996, *227*, 300-303.

Raju, P. V. S. N. P., Mandal, D. Study on the effect of free acidity and entrained TBP in UNPS on the quality of ADU powder. *Journal of Nuclear Materials* 2015, *464*, 346–355.

Ramana Rao, S. V. R., Lahiri, D., Anuradha, M., Murty, B. N., Palanki, B. An investigation on the effect of heating rate on the sintering of uranium dioxide. *Transactions of the Powder Metallurgy Association of India*, 2006, *32*, 100-113.

Ramm, E. J., Quass, H. T. Preparation and evaluation of an ADU derived ceramic grade UO_2 powder, *Australian Atomic Energy Commission AAEC-TM*-439, 1968.

Rhee, Y. W., Kim, D. J. Kim, J. H., Yang, J. H., Kim, K. S., Kang, K. W., Song, K. W. Fabrication of Sintered Annular Fuel Pellet for HANARO Irradiation Test. *Journal of Nuclear Science and Technology* 2010, *46*, 345-350.

Rhodes, W. W. Agglomerate and particle size effects on sintering of yttria stabilized zirconia. *Journal of the American Ceramic Society* 1981, *64*, 19-22.

Rodriguez, S. A., Hernandez, R. R., Garcia, C. R. Ma., Macias, B. L. R. Thermal decomposition of ammonium uranate at atmospheric pressure: infrared study. *Vibrational Spectroscopy* 1995, *9*, 215-219.

Rothman, A. B., Grazyk, D. G. Ammonium poly uranate precipitation from fluoride solutions: an experimental study of the UO_3-HF-NH_3-H_2O system as applied in UF_6 processing. *Nuclear Technology* 2009, *167*, 410-420.

Runfors, U., Schonberg, N., Kiessling, R. The sintering of uranium dioxide, *2nd United Nations International Conference on the Peaceful Uses of Atomic Energy,* A/CONF.15/PA42, Sweden, 2 June 1958.

Seneda, J. A., Figueiredo, F. F., Abrao, A., Carvalho, F. M. S., Frajndlich, E. U. C. Recovery of uranium from the filtrate of ammonium diuranate prepared from uranium hexafluoride. *Journal of Alloys and Compounds* 2001, *323-324*. 838-841.

Song, K. W., Kim, K. S., Kang, K. W., Kim, Y. M., Yang, J. H., Jung, Y. H. Effect of TiO_2 on sintering behavior of mixed UO_2 and U_3O_8 powder compacts. *Journal of the Korean Nuclear Society* 1999, *31*, 455-464.

Steeper, T. J., Zink, J. C. Particle-size distribution of ammonium diuranate precipitate, Proceedings *of the Oklahoma Academy of Science*s, 1974, 83-87.

Stehle, H., Assmann, H., Wunderlich, F. Uranium dioxide properties for LWR fuel rods. *Nuclear Engineering and Design* 1975, *33*, 230-260.

Tel, H., Eral, M. Investigation of production conditions and powder properties of AUC. *Journal of Nuclear Materials* 1996, *231*, 165-169.

Thermodynamic and transport properties of uranium dioxide, IAEA Technical Report Series No. 39, 1965, p.4.

Thompson, R. A. Mechanics of Powder Pressing; I, Model for powder densification. *American Ceramic Society Bulletin* 1981, *60*, 237–243.

Thompson, R. A. Mechanics of Powder Pressing; II, Finite Element Analysis. *American Ceramic Society Bulletin*, 1981, *60*, 244–247.

Thompson, R. A. Mechanics of Powder Pressing; III, Model for the Green Strength of Pressed Powders. *American Ceramic Society Bulletin*, 1981, *60*, 248–251.

Tousey, M. D. "Optimal Tablet Press Operation," *Pharmaceutical Technology*, 2002 http://techceuticals.com/wp-content/uploads/2016/07/Article-Optimal-Tablet-Press-Operation-Machine-vs. Granulation.pdf.

Vanderlaan, J. "Automated inspection of sintered uranium fuel pellets," *MS Thesis,* The University of Ontario Institute of Technology, Nov 2011.

Vasilos, T., Rhodes, W. "Solids processing of fine grain size ceramics" in: *Characterization of Ceramic Materials, Proceedings of the First US-Japan Seminar on Basic Science of Ceramics*, S. Saito, Y. Kotera & S. Somiya (Eds.), Tokyo and Kyoto, 24 February - 3 March, 1969, pp. 93-140.

Vold, M. J. The sediment volume in dilute dispersions of spherical particles. *Journal of Physical Chemistry*. 1960, *64*, 1616-1619.

Wang, J. W., Ewing, R. C., Becker, U. Average structure and local configuration of excess oxygen in UO_{2+x}. *Scientific Reports* 2014, 1-5.

Whittemore, O. J., Varela, J. A. "Pore growth and shrinkage during sintering," *4th International Symposium on Science and Technology of Sintering,* Tokyo, Japan 4-6 November 1987.

Wood, G. A. *"A simple model for the sintering behavior of IDR* UO_2 *powder,"* BNFL, STD-CP-06-34, 1999.

Woolfrey, J. L. The preparation of UO_2 powder: effect of ammonium uranate properties. *Journal of Nuclear Materials* 1978, *74*, 123-131.

Wu, C. Y., Hancock, B. C., Mills, A., Bentham, A. C., Best, S. M., Elliott, J. A. Numerical and experimental investigation of capping mechanisms during pharmaceutical tablet compaction. *Powder Technology* 2008, *181*, 121–129.

Yang, J. H., Kim, J. H., Rhee, Y. W. Improvement of UO_2 Pellet Properties by Controlling the Powder Morphology of Recycled U_3O_8 Powder. *Journal of Nuclear Science and Technology* 2008, *45*, 1150-1154.

Yang, J. H., Song, K. W., Lee, Y. W., Kim, J. H., Kang, K. W., Kim, K. S., Jung, Y. H. Microwave process for sintering of uranium dioxide. *Journal of Nuclear Materials* 2004, *325*, 210–216.

Yato, T., Onoue, T., Tanaka, H. *UO_2 pellet fabrication process, Mitsubishi Patent US* 4,882,100 November 21, 1989.

Yeap, S. P. Review: Permanent agglomerates in powdered nanoparticles: Formation and future prospects, *Powder Technology* 2018, *323*, 51–59.

Yoo, H., Lee, S. S., Kim, J., Song, K. Effect of UO_{2+x} Powders Produced at Different Oxidation Temperatures on the Properties of Pellet. *Journal of the Korean Ceramic Society* 2003, *40*, 410-414.

Zheng, J., Reed, J. S. Particle and granule parameters affecting compaction efficiency in dry pressing. *Journal of American Ceramic Society* 1988, *71*, C456–C458.

Section B

Energetic Interactions

> So I could see that if I were going to conduct any experiments of uranium hexafluoride, I had to get the uranium hexafluoride.
> This could not be bought. Nobody made it, so I must make some… and I devised a scheme for making uranium hexafluoride in good quantities by making first UF_4 and then converting this to UF_6. So I personally made the first several hundred pounds of uranium hexafluoride that was made in this country.

- *Philip Abelson*

Puzzle 2

Uranium Hexafluoride Interactions

by rockratzero

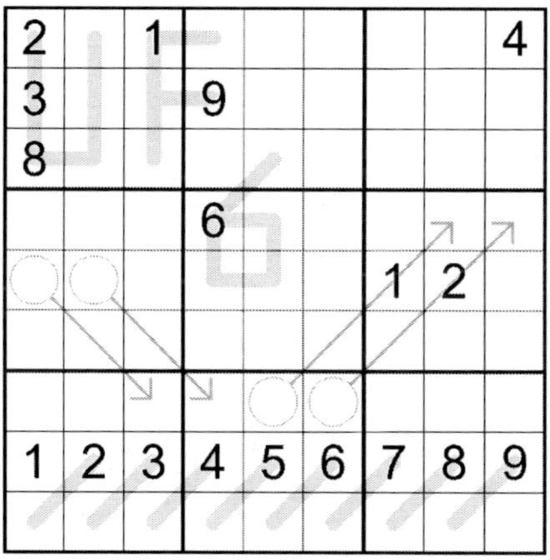

Rules

Place the digits 1-9 in each row, column, and 3x3 region without repeating.

Digits on an arrow line must sum to the value of the digit in the attached circle.

Digits on the gray lines alternate between odd and even numbers.

Link to SudokuPad online: https://tinyurl.com/rrz-uranium2.

Chapter 3

Chemistry of Organic Surface Reactions with Uranium Hexafluoride

Kristin K. Anderson[1,2]
Cari A. Launiere[3]
and William A. Alexander[1,*]
[1]Department of Chemistry, The University of Memphis, Memphis, TN, USA
[2]Department of Chemistry, Virginia Tech, Blacksburg, VA, USA
[3]Argonne National Laboratory Chemical and Fuel Cycle Technologies Division, Argonne, Il, USA

Abstract

Despite the importance of UF_6 in the nuclear industry, scant data are available on the interaction of UF_6 with organic surfaces. To begin to fill this knowledge gap, we have examined, using *in situ* and *ex situ* RAIRS spectroscopy, the stability and reactivity of self-assembled monolayer (SAM) surfaces bearing different organic functional groups. Reflection-Absorption InfraRed Spectroscopy (RAIRS) was used both to characterize the model organic surfaces and to monitor reaction progress (if any) during uranium hexafluoride exposure. The results presented here indicate that hexafluoride exposure on amino- and methyl-terminated SAMs results in physisorped UF_6 surface layers, or reactive deposition of a species like UO_2F_2, while fluorinated SAMs are more resistant to hexafluoride attack. Only the NH_2-SAM showed compelling evidence for extensive reaction with the SAM chains themselves. RAIRS characterization showed successful consistent creation of SAM-functionalized surfaces. These prepared surfaces were exposed to

* Corresponding Author's Email: w.alexander@memphis.edu

In: The Chemistry of Uranium
Editor: William A. Alexander
ISBN: 979-8-89530-802-8

gaseous WF_6 or UF_6 in a custom reaction chamber and any reaction progress monitored with *in situ* RAIRS. Based on the analysis herein, and with extrapolation from WF_6 exposure data, it appears that the fluorinated SAM is resistant to chemical attack by hexafluorides, while amino-terminated SAMs are degraded. All surfaces studied show a relatively dramatic shift in their backbone stretching frequencies, presumably from initial absorption of UF_6 perturbing the overall SAM order. Further experiments in the future would help confirm whether the changes seen in the surfaces are true reactions or a physical effect of the adsorption/desorption process.

1. Introduction

This chapter discusses a chemistry investigation that was conducted as part of a larger effort to determine the feasibility of gas-phase UF_6 alpha spectroscopy for enrichment measurements. Gas-phase UF_6 alpha spectroscopy was investigated as a candidate technique to complement the On-line Enrichment Monitor (OLEM) for near-real time enrichment verification in gas centrifuge plants. The OLEM directly measures only ^{235}U and relies on several assumptions as well as operator temperature and pressure measurements to estimate the total uranium [1]. Alternatively, alpha spectroscopy can be used to directly measure all three isotopes of interest without relying on any of the assumptions required by the OLEM. This would make alpha spectroscopy a useful tool for validating OLEM measurements if it could be adapted for at-line use. Alpha detectors are robust, inexpensive, compact, and use little energy, thus making them a good candidate for process deployment. Generally, the main limitation of alpha spectroscopy is that it usually requires extensive sample preparation to produce homogeneously thin sources using techniques such as vacuum sublimation or electrodeposition. However, because UF_6 gas is already homogeneous, the depth (in molecular layers) of a gas-phase UF_6 alpha spectrometry sample can be controlled easily and precisely by controlling the thickness and partial pressure of the gas column that a detector is exposed to. Therefore, it should be feasible to perform alpha spectrometry on gas phase samples without the need for any sample processing.

The key technical limitation for this approach is the ability of the detector to withstand the harsh environment of highly reactive UF_6 process gas. Alpha detector surfaces are silicon, which is highly susceptible to damage by UF_6 and HF gas, so a passivating coating is required. Materials commonly used in

UF_6 processes, even those considered to be relatively inert, will degrade over time due to reaction with UF_6 and HF. However, these reactions involve complex porous surface structures with significant quantities of a wide variety of impurities [2, 3]. For this report, high purity thin-film coatings known to be inert to UF_6 gas (gold and AlF_3) were investigated.

Gold is one of the most inert materials in the world. It is known to be non-reactive with HF and with UF_6 up to 400°C [4, 5]. Gold apparatuses have been used in UF_6 analytical applications, and thin-film gold (50 nm) has been used in thermochromatographic alpha particle decay measurements with silicon PIN-photodiodes [6]. Gold coatings can be applied to silicon surfaces by the low-cost process of physical vapor deposition (PVD), and complete gold coverage can be achieved with layers as thin as 6 nm. Alpha spectroscopy modeling results indicated that alpha particle attenuation in such a thin gold layer would be minimal [7]. These properties make gold a candidate material for passivating silicon detector surfaces. Furthermore, gold surfaces can be further modified via the deposition of organic thin films.

Aluminum is a common material used for UF_6 process pipes because a passivating layer of AlF_3 is formed upon exposure of aluminum to UF_6. The AlF_3 passivating layer formed on an aluminum alloy process pipe is thick and porous due to reactions involving oxides, impurities, and HF. However, it is possible to directly deposit an ultra-pure and very thin (3 nm) layer of AlF_3 onto a silicon surface by atomic layer deposition (ALD). Because AlF_3 has one-third the atomic number density of Au, and can be deposited as a very thin film, it is expected to have negligible effects on peak broadening when applied as an alpha detector coating.

The available literature on uranium hexafluoride gas-surface interactions covers materials relevant to the enrichment and chemical processing of UF_6. These materials include metals (carbon steels, stainless steels, Monel®, nickel, copper, aluminum, and gold), materials used in seal connections (butyl rubber, fluorinated thermoplastics, and fluorinated elastomers), and materials used as sorbents and filters (activated carbon, Al_2O_3, NaF, KF, CaF_2 and MgF_2) [2-5, 8, 9]. Partially because of the abundance of depleted UF_6 that is produced from the enrichment process, many previous studies report the use of UF_6 for the oxidation or fluorination of various organic compounds [3, 4, 8-20]. However, while some of these studies offer some insight into product formation through post-reaction product analysis [4, 15-21], to the best of our knowledge, no previous works have studied the *in situ* gas phase reactions of UF_6 with different organic materials. In this work, we exposed different surfaces first to tungsten hexafluoride (WF_6) gas, which may act as a simulant gas for UF_6, and compared the true behavior of UF_6 on the same surfaces. WF_6 has the same molecular geometry and is thought to display similar reactivity to UF_6. While WF_6 has been shown to be a milder oxidizer, there have been conflicting reports on the strength of UF_6 as an oxidizer [16, 17, 22]. The conflicting data between reports appears to stem from varying reaction conditions. By spectroscopically monitoring gas-surface reactions, we hope to offer more insight into unexplored reaction pathways.

To this end, within this chapter we describe the interaction of UF_6 and WF_6 with a special class of thin-film organic materials that are formed through self-assembly by deposition of alkanethiol precursors onto gold. Self-assembled monolayers (SAMs) act as a good model organic surface and have been shown previously to offer insight into how gases react at organic surface interfaces [23-26]. The sulfur atom of the thiol end group chemisorbs to gold in well-defined binding sites. A long chain hydrocarbon backbone connects the sulfur end group to a functional head group. Cooperative van der Waals interactions between the methylene units of the chemisorbed alkyl chains help create a densely packed, well-ordered monolayer known as a self-assembled monolayer or SAM (see Figure 1). Because of this ordering, the functional head group is exposed precisely at the gas-surface interface. By optimizing the interfacial functionality through the selection of various SAM chain head groups, the surfaces can be tuned to achieve different UF_6 gas interaction modalities.

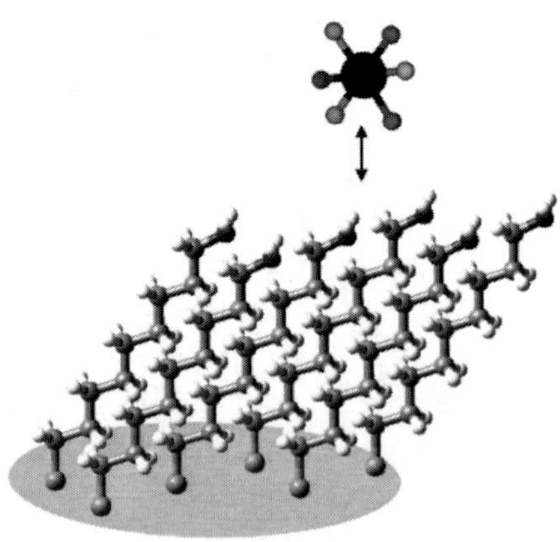

Figure 1. Schematic of a UF_6 molecule interacting with an NH_2-terminated alkanethiolate self-assembled monolayer on a gold surface.

UF_6 interaction with three different SAM surfaces (hydrocarbon, fluorocarbon, and amine-terminated) and two non-SAM surfaces (Au and AlF_3) were studied using a custom UF_6 materials testing apparatus of our design at Argonne National Laboratory. The system featured in-situ analysis capabilities for analyzing surfaces before, during, and after UF_6 exposure. These in-situ analysis capabilities include reflection-absorption infrared spectroscopy (RAIRS), transmission Fourier transform infrared spectroscopy (FTIR), and a quartz crystal microbalance (QCM). An on-instrument residual gas analyzer (RGA) was used to monitor residual gas impurities and volatile reaction products. Details regarding materials preparation and analysis methods are given at the end of the chapter for the interested reader.

2. Results and Discussion

2.1. WF_6 Pulsed Exposure Quartz Crystal Microbalance Studies

An atomic layer deposition (ALD) instrument was used for pulsed-exposure WF_6 QCM studies to leverage the rapid response flow control to observe reaction and adsorption kinetics. Figure 2 shows representative results from the pulsed WF_6 exposure studies. All surfaces demonstrated reproducible results except the fluorocarbon SAM, which behaved like a non-reactive surface in most studies (lower, darker CF_3 trace) but behaved like a reactive surface for about 20% of samples (higher, lighter CF_3 trace). Fluorocarbon SAMs have lower packing density and weaker interchain bonding networks than hydrocarbon chains so this could be due to flaws in the prepared SAM that exposed adsorption or reaction sites within the SAM monomer chains. Of the hydrocarbon SAM surfaces, those with head groups that chemically react with WF_6 (N-methylimidazole-terminated and amine-terminated SAMs) exhibited deposition of approximately one monolayer of material (approximately 270 ng/cm^2) within one to six gas pulses. Conversely, the hydrocarbon SAM had no detectable mass deposition following the first pressure pulse. One might expect the hydrocarbon SAM surface to be highly reactive with the hexafluoride gas because of the notion that liquid phase hydrocarbons undergo rapid, sometimes explosive, fluorination reactions upon exposure to liquid UF_6 [27]. However, it appears that steric hindrance in the tightly-packed SAM CH_3 head groups provides significant resistance to fluorination attack. The surface is not entirely unscathed, as slow deposition onto the hydrocarbon surface does occur over time. This could be a result of reactions with incoming water contaminants, as discussed below. The inert gold surface had significantly less deposition than the reactive services, but significantly more than the AlF_3 surface. Importantly, the un-modified gold, hydrocarbon SAM, fluorocarbon SAM, and AlF_3 surfaces all appeared to have less than a monolayer of deposited material by the end of the experiment.

Figure 3 shows the mass deposition broken down into its irreversible (Figure 3A) and reversible (Figure 3B) components. Irreversibly deposited mass consists of W, F, O, or H atoms that remain on the surface following each pulse after having undergone chemical reactions with the surface or species adsorbed on the surface. These reactions could occur with the SAM surface groups, with the hydrocarbon SAM chains (fluorination reactions), with residual surface water, or with contaminant water entering with the cover gas. Except for the hydrocarbon SAM, the SAM surfaces tended to have higher initial irreversible mass deposition rates, particularly those with reactive nitrogen-containing head groups. Irreversible deposition on the SAM surfaces converged to a constant deposition rate within 15 pulses. The Au and AlF_3 surfaces did not converge with the SAMs and had lower overall rates of deposition. XPS data (not shown) also support the conclusion that all surfaces have WO_3 deposition as a result of reactions with water, although the total WO_3 content appeared to vary in correlation with the deposition rates observed in these QCM studies. The effective roughness of the surface appears to be a key component in the observed distribution of the deposited-mass-per-pulse measurements seen here. The AlF_3 surface is expected to have the lowest surface roughness because it is produced by atomic layer deposition. The Au surfaces are produced by physical vapor deposition, which results in higher surface roughness and more defects and grain boundaries than AlF_3. A rough surface provides a larger surface area and a wider diversity of potential adsorption sites to nucleate reactive deposition. The SAM coatings (which are applied on top of the Au surfaces) are expected to add additional effective surface roughness – while SAM surface roughness measurements generally follow that of the underlying substrate, the additionally flexibility of the semi-crystalline organic SAM allows impinging gas species an increased diversity of adsorption sites. Additionally, depending on the functional group species, intrachain hydrogen-bonding can disrupt the otherwise well-ordered close-packed exterior of the SAM to allow access to the interior of the monolayer.

Unexpectedly, the fluorocarbon SAM and un-modified Au surfaces (which should be inert to WF_6) had the second highest initial irreversible deposition rates behind the reactive nitrogen-containing headgroups. This mass deposition is likely the result of reactions between residual surface water and WF_6, which adsorbs strongly on these surfaces as shown in Figure 3B. Also unexpectedly, the hydrocarbon SAM had the lowest initial irreversible deposition rate of any surface tested. This was the result of the low reversible deposition rate observed in Figure 3B. Recall that only rigidly adsorbed molecules can have their mass accurately measured by QCM, so QCM characterization of weak adsorption is more qualitative in nature. It is possible that the low masses in Figure 3B represent weakly adsorbed monolayers rather than partial monolayers. The observed trends are in line with modeling results of UF_6-functional group dimers (See Chapter 4), which predicted intermolecular binding energies of ~2, 13, and 15 kcal/mol for interactions between UF_6 and the hydrocarbon, imidazole, and amine surface groups, respectively [29]. The fluorocarbon SAM was an outlier with a calculated potential well depth of ~1 kcal/mol. The higher reversible adsorption than predicted for this surface may result from interactions with the monolayer interior or differences in the WF_6 and UF_6 chemistries.

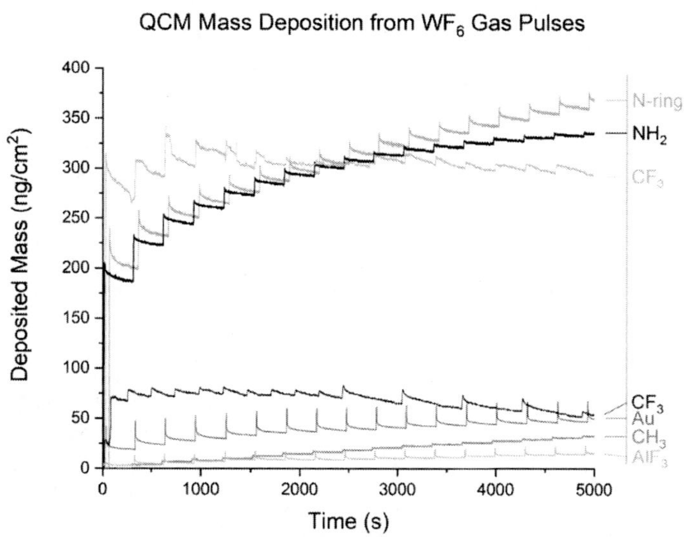

Figure 2. Mass deposition on surfaces following exposure to WF_6 gas. QCM measurements show mass deposition on AlF_3, Au, hydrocarbon SAM, fluorocarbon SAM, amine-terminated hydrocarbon SAM, and N-methylimidazole-terminated hydrocarbon SAM ("N-ring") surfaces upon exposure to 5 second pulses of WF_6 gas under 1 torr argon purge gas. The darker "CF_3" below is indicative of most experiments, while the lighter "CF_3" trace above shows behavior seen rarely (~20% of samples) and thought to be from poor surface preparation.

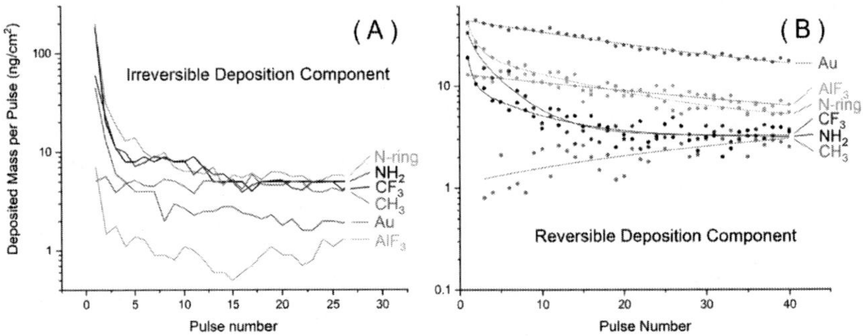

Figure 3. Mass deposition divided into (a) irreversible and (b) reversible components.

It should be noted that the QCM technique is very sensitive and the total permanent deposition on the inert surfaces was less than a single apparent monolayer at the end of the experiment. Therefore, the rate of deposition on these surfaces is incredibly slow. It was determined that the sub-ppm quantity of water in the ultra-dry cover gas is sufficient to cause this rate of deposition. It was clear that the amine-terminated surface is reactive with the hexafluoride gas. This makes this surface a useful control for comparison to the inert surfaces. The overall take away from these results is that inert thin-film coatings can provide protection against deposition during exposure to hexafluoride gas.

2.2. QCM Studies of UF_6 – Surface Exposure

WF_6 is a reasonably good surrogate for UF_6 in terms of physical and chemical properties [10]. However, UF_6 does have an order of magnitude lower vapor pressure than WF_6 and is a slightly stronger oxidizer. Therefore, experiments with UF_6 are essential to validate the trends observed in the WF_6 studies. Additional QCM studies were carried out in the new UF_6 exposure chamber, which uses UHV bake-out capabilities for water removal in lieu of the high-velocity inert gas purge used in the ALD system. The results below present the QCM response as the crystal frequency shift rather than attempting to convert the frequency shift to mass deposition because such a calibration would likely be inaccurate due to the varying adsorption strengths of UF_6 on the different inert surfaces. The deposition in these longer-term studies was generally greater than 99% irreversible, indicating that UO_2F_2 is being deposited due to reactions with water contaminants. UF_6 adsorption appeared to cause a five-times greater shift in frequency compared to WF_6 adsorption in this system (WF_6 data not shown) as may be expected from its lower vapor pressure. Since stronger adsorption has been shown to increase the WO_3 deposition rate, the inert surfaces were not expected to perform as well with UF_6 in the presence of water contaminants as they do with WF_6.

Figure 4 shows the immediate response of the coated QCM detectors upon exposure to 1 Torr UF_6 gas. As expected based on the WF_6 results, mass quickly deposited on the reactive amine surface but deposited more slowly on the inert surfaces. The Au and AlF_3 plots lines would overlap with the fluorocarbon plot line at this scale and are omitted for clarity. In contrast to the WF_6 experiments, deposition on the fluorocarbon surface occurred at a lower rate than the hydrocarbon surface. This could be a result of the subtle differences in chemistry between the two gas molecules, which are noted in the RAIRS discussion below. The differences in deposition extend to longer-term exposures as shown in Figure 5. For these experiments it appears that a significant amount of the UF_6 released into the chamber reacted to form UO_2F_2 on the chamber surfaces through reaction with small amounts of water contaminants that seep into the system through the various seals. There is a clear trend in the deposition rates and total deposition on the different coatings with $NH_2 > CH_3 > Au > CF_3 > AlF_3$.

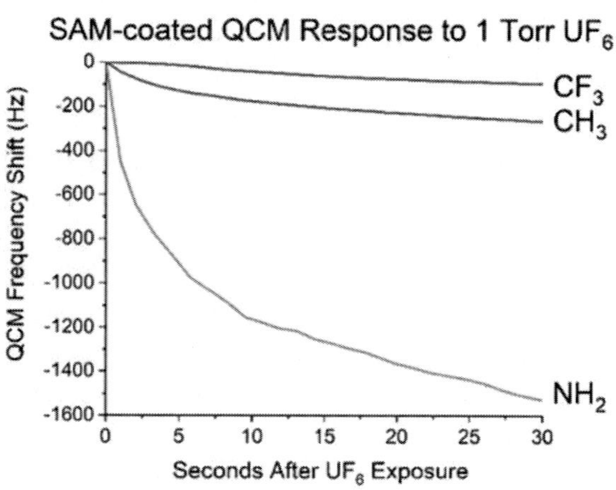

Figure 4. Mass deposition on surfaces upon exposure to UF_6 gas, immediate response.

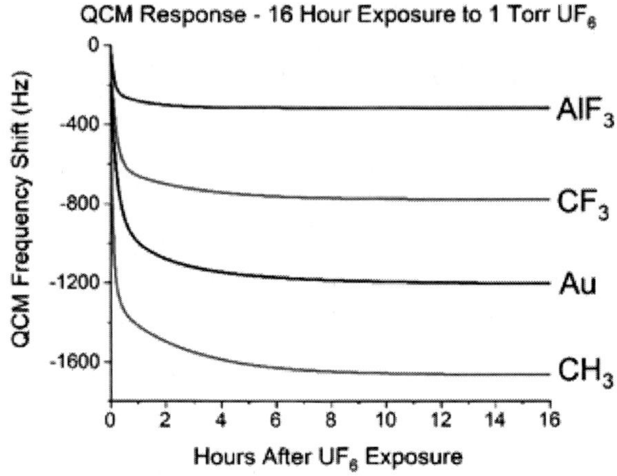

Figure 5. Mass deposition on surfaces upon exposure to UF_6 gas, 16-hour response.

The performance of the coatings in UF_6 appears to be heavily influenced by the same two factors observed in the WF_6 results. First, surface roughness appears to play a key role. Second, UF_6 adsorption strength appears to have an effect, since surfaces with higher UF_6 interaction potentials [29], have increased UO_2F_2 deposition relative to those with lower interaction potentials. Under a continuous flow of UF_6 - with a continuous source of water contaminants - this effect would likely be irrelevant since all surfaces would eventually behave the same after being converted to UO_2F_2. However, for a batch experiment, this property offers protection from deposition because UF_6 and water are preferentially reacted out on other chamber surfaces.

This effect could potentially be leveraged in batch sampling systems to prevent UO_2F_2 build-up on detector or window surfaces. It has become apparent that incoming UF_6 gas samples are likely to have small quantities of water contaminants in them as a result of poorly evacuated dead spaces and water ingress through seals in the gas delivery manifold. A potential solution to this problem would be to transition UF_6 gas samples through a "sacrificial" UHV-evacuated chamber containing a high-surface area sponge. The water in the samples would equilibrate with the dry surfaces and undergo surface reactions with UF_6 to form UO_2F_2. This would leave a dry gas sample to be transferred to a main experimental chamber. Using much larger alpha detector surface areas would allow shorter count times which would also help to minimize the amount of water entering the system through the sealing surfaces, as would an inert cover gas around the instrument.

2.3. RAIRS Studies of Surface Disordering and Reactions

For all the RAIRS spectra shown here, the top trace in the panel is the initial spectrum obtained under UHV before exposure to hexafluoride gas. The subsequent curves are difference spectra; they indicate any changes to the RAIRS signal relative to the initial curve. Positive peaks in these difference spectra indicate species appearing, while the loss of chemical species result in negative peaks. The gain or loss could be due to the actual appearance or disappearance of the species or due to the changes in the bond orientation with respect to the plane of the polarized light. For SAM surfaces, a negative peak in the CH_2 modes generally indicates a disordering of the SAM chains. Smaller peaks are generally indicative of reordering to accommodate adsorbing molecules, while larger shifts are generally the result of chemical reaction. Spectra were collected every 1 min at the beginning of each experiment, with longer time steps later in the experiment. Longer exposures and higher UF_6 concentrations were used for surfaces with slow reaction kinetics. To be able to better compare results between experiments, while accounting for various scan times and differences in exposure pressures, the spectra series are labelled at various points with the overall cumulative exposure in units of Langmuir. A Langmuir is equal to the pressure of gas (in 10^{-6} Torr) multiplied by time (in seconds). One Langmuir of gas exposure would, with unit sticking probability, represent the equivalent of one monolayer of gas adsorbing to the surface. Spectra were also obtained after exposure by removing the hexafluoride and pumping the chamber back to UHV - these post-exposure spectra are labelled 'vac.'

Initial tests were executed with WF_6 both to support instrument testing before switching to UF_6 and to better understand the differences between the gases (since the pulsed exposure QCM experiments and XPS experiments could only be carried out with WF_6). Figures 6A and 6B show RAIRS spectra for hydrocarbon SAMs exposed to WF_6 and UF_6, respectively. Increasing the WF_6 exposure up to 3 Torr over 230 minutes did not result in any discernable change to the CH_2 stretching mode peaks (data not shown), so the experiment was repeated at 8 Torr for 160 minutes (Figure 6A). Under these conditions a very small negative peak formed for the CH_2 asymmetric stretch mode indicating a very slight surface disordering. This response correlates well with the QCM data shown in Figure 2, where the initial response of the QCM to WF_6 exposure equates to a fraction of a percent of a rigidly adsorbed monolayer. It is possible that a partial or even a full monolayer of WF_6 formed on the hydrocarbon surface shortly after exposure, but the interaction was not enough to disrupt the tightly packed SAM chains.

Initial exposure to UF_6 had a more significant impact on the hydrocarbon SAM surface, as observed in Figure 6B for the 1 and 2 Torr exposure. Disruption of the asymmetric CH_2 stretching peak occurred upon exposure to just 0.01 Torr UF_6 within the first minute of exposure (data not shown). Further experiments showed that, as the initial exposure pressure was increased to 0.05 Torr, 0.1 Torr, and 1 Torr (1 Torr shown in Figure 6B), larger negative peaks are observed at the modes associated with CH_2 and CH_3 stretching. The increasing response of the SAM surface with gas pressure aligns with typical isotherm behavior, indicating a simple surface adsorption mechanism where the chains rapidly reorder to accommodate the adsorbing gas. It is apparent that UF_6 adsorbed more strongly on the hydrocarbon surfaces than WF_6 did, which likely led to the greater mass deposition (relative to the fluorocarbon surface) observed in the UF_6 experiments.

Further disruption of the hydrocarbon SAM chains is not observable following the initial reordering upon UF_6 exposure and no fluorine was detected in the XPS scans of the WF_6 exposed hydrocarbon SAM. This indicates, but is not conclusive evidence, that fluorination reactions of the tethered SAM chains are not occurring during UF_6 exposure. In preliminary non-reactive molecular dynamics simulations of this experiment (not reported here), the average maximal attractive potential experienced when the UF_6 interacts with the hydrocarbon SAM surface was only 6 kcal/mol. However, it was seen that UF_6 can experience attraction potentials as high as ~25 kcal/mol with the SAM surface. These maximal attractive potentials all occurred in deeply penetrating trajectories, which are likely to participate in chemical reaction processes such as fluorination of carbons in the SAM backbone. While this phenomenon was not readily apparent in these experimental results, modeling results indicate that SAM disordering due to UF_6 penetration should be considered as a possibility during long-term exposure. We did observe some spectroscopic evidence of gas-phase perfluorocarbon production, characterized especially by peaks centered at 1272 and 949 cm^{-1}, which was identified in a previous study as an impurity peak from UF_6 reacting and breaking down Viton (a perfluorocarbon polymer) O-rings in the system [30] – our system has no such components. When UF_6 is exposed to the hydrocarbon surface, this signal grows over time; we believe that some SAM chains are oxidized and cleaved from the surface at the sulfur head group, presumably at defect sites and grain boundaries. These cleaved chains can subsequently be fluorinated by UF_6 as free hydrocarbon chains in the gas phase, although we are not able to confirm this. Overall, the data indicate that the hydrocarbon SAM surface does not react with either WF_6 of UF_6, and the main contributor to mass deposition is reaction with surface water contaminants.

Figure 6C shows RAIRS spectra for amine-terminated SAM surfaces exposed to 0.05 Torr WF_6 for 200 minutes followed by 0.25 Torr for 200 minutes. In the WF_6 exposed sample, the CH_2 modes are seen to be slowly decreasing over time, indicating mass loss or degradation of the surface ordering. Based on the XPS data it is apparent that fluorination reactions occur on this surface, although it is unclear if it is the nitrogen, carbons, or both that are fluorinated. The amine-terminated SAM only has 10 methylene units in its linker chain, while the hydrocarbon SAM consists of a 17-methylene chain linker. The longer chain length in the hydrocarbon SAM serves to stabilize and order the surface, which results in tighter packing. The shorter amine-terminated SAM therefore likely has less resistance to WF_6 entering the SAM interior and fluorinating the chains compared to the hydrocarbon SAM. Molecular dynamics simulations show that the amine groups form a shifting hydrogen bonding network which may open breaks in the packing to expose the SAM chains for attack [26, 31]. Furthermore, the amine-terminated SAM had the highest fraction of penetrating trajectories of all the models. The amine-terminated SAM surface behaved similarly in UF_6, except the chain disruption occurred much more quickly for the same concentration of UF_6 (0.05 Torr) as shown in Figure 6D. Unlike the hydrocarbon surfaces, the persistence of a longer-term process can be seen clearly in the amine data. The apparent mechanism follows a quick surface re-ordering, followed by slower, longer scale process that likely involves irreversible reactions with the SAM chains.

Figure 7A shows RAIRS spectra for the fluorocarbon SAM surface under exposure to 6 Torr WF_6 for 280 minutes. Here the fluorocarbon stretching region is plotted, analogous to the hydrocarbon region shown in Figure 6. While no major positive or negative peaks are seen appearing in the fluorocarbon mode regions for the WF_6 exposed surface, the results are not fully conclusive due to the proximity of water peaks in the spectrum. The series of peaks from ~1325 cm^{-1} to 1500 cm^{-1} are from increasing gas-phase water vapor contamination in the beam path, likely introduced from the hexafluoride source lines. The water in Figure 7A must be largely due to water contaminants in the chamber itself as evidenced by the fact that the peaks do not disappear upon chamber evacuation. The strong positive peaks in the difference spectra indicate that water ingress is likely coming through the seals in the hexafluoride gas manifold chamber.

For the UF_6 experiments, the procedures were changed to evacuate the manifold gas line to UHV to minimize moisture contamination from this source. In these experiments (Figure 7D and 7E), the water in the chamber tended to decrease over time. This provides further evidence to support the assertion that mass deposition on surfaces is due to UF_6 reacting with surface water to form UO_2F_2. It is also an indication of slow gas-phase UF_6 water reaction kinetics at low water concentrations.

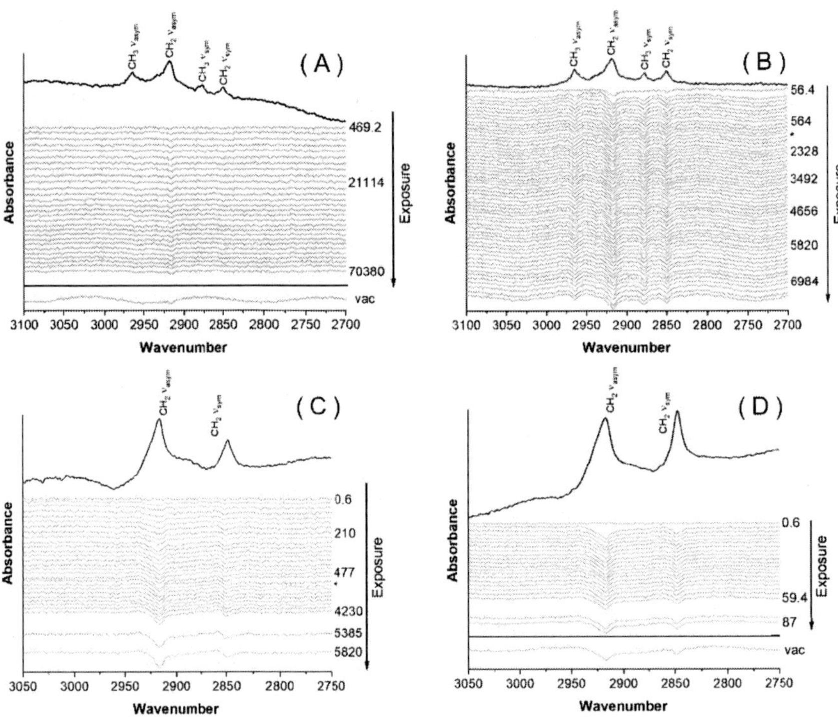

Figure 6. RAIRS monitoring of hydrocarbon and amine SAM surfaces during hexafluoride exposure. (A) Hydrocarbon SAM, 8 Torr WF_6; (B) Hydrocarbon SAM, 1 and 2 Torr UF_6; (C) Amine-terminated SAM, 0.05 and 0.25 Torr WF_6; (D) Amine-terminated SAM, 0.05 Torr UF_6. Exposure is reported in Langmuir/10^{-6}. *denotes point of pressure increase of UF_6 addition.

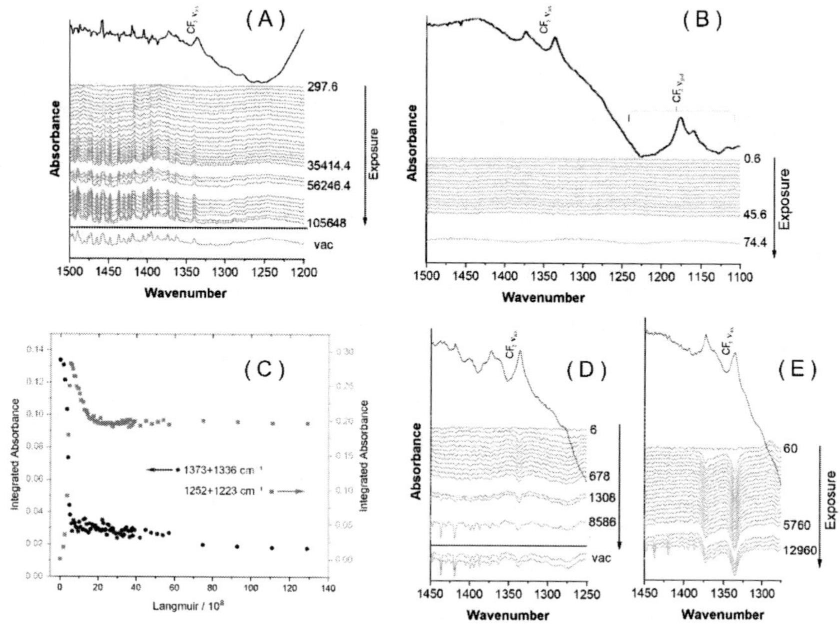

Figure 7. RAIRS monitoring of fluorocarbon SAM surfaces during hexafluoride exposure. (A) 6 Torr WF_6; (B) 0.01 Torr UF_6; (C) Integrated absorbance the CF modes of the pristine SAM and the new CF modes that appear after exposure of 1.00 torr UF_6 vs exposure; (D) 0.1 Torr UF_6; (E) 1.0 Torr UF_6. Exposure is reported in Langmuir/10^{-6}.

In contrast to WF_6, surface disordering could be seen upon UF_6 exposure to the fluorocarbon SAM. The labelled peaks at 1373 and 1336 cm^{-1} are bands assigned to the vibrational modes for CF_2 that have a transition dipole parallel to the perfluorohelical axis, referred to as axial bands, CF_2 ν_{ax} [32, 33]. This mode shows maximal signal when the SAM chain is aligned with the surface normal – i.e., the chain is not tilted. Between ~1100 to 1300 cm^{-1} are expected a number of overlapping bands attributable to the vibrational stretches for CF_2 that have transition dipoles perpendicular to the perfluorohelical axis, CF_2 ν_{pd}. Within these bands, peaks at 1252 and 1223 cm^{-1} are seen when the fluorocarbon chain tilts away from the surface normal [32]. The lack of these diagnostic CF_2 ν_{pd} peaks indicates that the perfluoro chains that make up the monolayer initially are all well-ordered and aligned with the surface normal orientation. Clearly no disordering or surface reaction is seen at the low 0.01 Torr UF6 exposure represented in Figure 7B, but increasing the pressure to 0.1 or 1 Torr (Figure 7D and 7E, respectively), we see decreases in the CF_2 ν_{ax}

modes, and new peaks building in at 1252 and 1223 cm^{-1} in the CF_2 ν_{pd} region (not shown). Figure 7C shows how the CF_2 modes of the SAM change over time. The figure shows that as the CF_2 ν_{ax} modes decrease, the new modes at 1252 and 1223 cm^{-1} increase over time. Some of the decrease in CF_2 ν_{ax} modes is due to chains being cleaved from the surface via reaction at the sulfur head group, predominately at defect sites and grain boundaries in the layer. However, most of the chains must remain attached to the surface, tilting away from surface normal, as evidenced by the growth of the CF_2 ν_{pd} peaks. This simultaneous decrease and increase in the parallel and perpendicular modes, respectively, indicate that upon exposure to UF_6 gas, the surface undergoes substantial rearrangement. As fully fluorinated species are known to be resistant to UF_6 fluorination, so any reaction must come from the Au-S-C bond of the thiol [19, 33]. While the perfluoroalkane chain is bulkier than the hydrocarbon chains, the sulfur heads are spaced further apart to accommodate (i.e., ~5.78 Å for fluorocarbon SAM vs 4.98 Å for hydrocarbon SAM) [34, 35]. While the fluorocarbon surface is more densely packed, all SAM surfaces contain defects that make the sulfur groups accessible to UF_6 and the larger distance between Au-S moieties likely helps facilitate UF_6 attack at these defect sites.

Overall, the fluorocarbon surfaces behaved similarly to the hydrocarbon surfaces upon exposure to different UF_6 concentrations. The initial chain disruption occurred within the first 2 minutes and increased with increasing exposure pressure from 0.01 to 0.1 to 1.0 Torr (Figures 7B, 7D, and 7E, respectively). Unlike the hydrocarbon SAMs, initial chain disruption was not discernable at the lowest exposure pressure of 0.01 Torr (Figure 7B). It is expected that the SAM would be accommodating a quickly formed (perhaps partial) monolayer of UF_6 molecules on the surface, but that this UF_6 layer does not appear to perturb the underlying SAM surface. It seems that the extra density and increased inertial mass of the fluorocarbon SAM compared to the other two surfaces makes it able to withstand the initial SAM re-ordering. This likely results in weaker adsorption of UF_6 on the fluorocarbon surface versus the hydrocarbon surface. This assertion is supported by the QCM results in Figure 4 which show quicker deposition or stronger adsorption of UF_6 on the hydrocarbon surface versus the fluorocarbon surface and by results in Figure 5 which show more total deposition on the hydrocarbon surface at the end of the experiment. The assertion is also supported by preliminary molecular dynamics simulations (not shown), which indicate an average interaction potential of 2 kcal/mol for the fluorocarbon surface versus 6 kcal/mol for the hydrocarbon surface. At higher pressures (Figure 7D and 7E) the fluorocarbon SAM still resists any initial change, but over time, the chains start to show disorder as the surface responds to chain loss at defect sites and depositing UO_2F_2.

The results with the SAM surfaces have shown that these surfaces may be useful for organic synthesis reactions that use UF_6. Organic synthesis has been identified as a potential use path for depleted UF_6 because UF_6 is a powerful reactant that is available in large quantities at low to no cost. However, the reactions of interest often have poor yields and large numbers of reaction products when carried out in the bulk phase. Surface reactions on sorbents have been shown to improve yield [9]. It is possible that the precise orientation of SAM surface functionalities and the tunable UF_6 interaction properties demonstrated herein could be leveraged to achieve orders of magnitude yield improvements in these types of reactions.

2.4. Reaction Products, Water Contamination and UO_2F_2 Formation

Based on the cumulative results of this project, it was determined that water contaminants entering through valve and fitting seals posed a major challenge for gas phase alpha spectroscopy, which is highly sensitive to alpha particle attenuation and peak tailing as a result of material deposition on the detector surface. Some sources report that UF_6 gas is completely anhydrous due to rapid water reaction kinetics [8, 36]. For example an NRC application by AREVA states that "UF_6 reacts with water so rapidly that HF remains anhydrous when in the presence of UF_6" [8]. It is possible that long-operating industrial enrichment facilities have a build-up of UF_6-water reaction products that create a tortuous path with lots of surface area such that much of the water is removed through surface reactions before it can enter the bulk phase. However, to our knowledge, no data on the water content of UF_6 process gas is available. Furthermore, the (sometimes significant) water content in UF_6 transport cylinders as a result of the formation of UO_2F_2 complexes with H_2O and HF has been well-documented [37].

The trimolecular reaction between UF_6 and water is known to be slow in the gas phase at low water concentrations [38, 39], which allows water to be present in UF_6 gas for extended periods of time. However, the reaction is catalyzed by surfaces [40], allowing surface reactions to continue after gas-phase reactions have become kinetically unfavorable. This was observed in the present studies, where water entered the sealed chamber and reacted with QCM and alpha detector surfaces in the center of chamber. There was no evidence of molecular clusters stabilizing water molecules in the gas phase in our experiments, but there are still many measurements that need to be performed to fully understand these systems.

Previous attempts to characterize UF_6 systems have often over-simplified the system by ignoring water and other contaminants. However, because this project focused on surface science (which is highly sensitive to trace impurities) highly sensitive instruments such as the residual gas analyzer (RGA) were employed. Figure 8 shows the RGA background spectra for a chamber evacuated to UHV following a UF_6 exposure experiment (dark), and for the head space of the UF_6 gas tank (light). To achieve the second measurement the chamber and gas line were first evacuated to 10^{-8} Torr, then the head space of the UF_6 tank was opened to the gas line for 8 hours before injecting a small quantity of gas into the chamber from the equilibrated line. As shown in Figure 8, the UHV evacuated chamber contained small amounts of H_2O (16 and 18 amu), F (19 amu), HF (20 amu), and N_2 (28 amu). These species are indicative of air contamination and the formation of UO_2F_2 and HF. UF_6 is too large for detection in this instrument, but large fluorine peaks were generally considered to be the result of residual UF_6. The water peaks in the RGA spectra never dropped below the limit of detection of the instrument throughout the entire project, which highlights the ubiquity of water, even in hexafluoride and UHV systems.

The plot highlights the complexity of the gas system in the UF_6 tank and gas manifold. The ratios of the constituents are not representative because much of the lower vapor pressure UF_6 would have adsorbed on the chamber walls before reaching the RGA. This gas system is a complex mixture of moisture and other air constituents, which likely leak in through the valves and fittings over time, and hydrocarbon impurities. These impurities are in trace quantities that will not affect bulk scale gas-phase systems over the short-term but will have a significant impact over the short-term on surface sensitive systems. For this reason, surface reactions and impurities should be factored into experimental design for experiments involving surface science or involving precision measurements that could be affected by surface reactions. Furthermore, these impurities can have a significant impact in bulk scale systems over the long-term as observed in the clogging of UF_6 process pipes near sites of water ingress. For this reason, water impurities should be carefully considered and controlled for in systems intended for long-term deployment at enrichment plants. It should be noted that there was no visible clouding or build-up of material on the exposure chamber windows throughout the whole project. The UO_2F_2 deposition was only observed through the sensitive instrumentation. However, by carefully measuring the slow rate of deposition in these systems, deposition over long time scales can be estimated and used to determine design parameters and predict performance.

With it established that in our system the hexafluoride gas source brings with it atmospheric and other contaminants, we now reexamine a few features of the RAIRS experiments in more detail. Figure 9 displays a few regions of interest in the spectra from UF_6 exposure on the hydrocarbon SAM. First we can examine the adsorption of UF_6 gas on the surface. Various UF_6 vibrational bands can be seen in the RAIRS spectra during exposure. The principle ν_3 UF_6 mode, centered at 625 cm^{-1}, shows a near constant signal. Additional combination bands for UF_6 can be seen in Figure 9A and 9B around 1291 and 670 cm^{-1} [41], confirming the constant UF_6 signal. These signals disappear after the chamber is evacuated at the end of the experiment, as expected.

Three other gas-phase peaks are of interest in Figure 9A and 9C. First is a peak that appears at 1031 cm^{-1} and grows in over time; this peak belongs to gas-phase SiF_4, which is produced due to UF_6 etching the glass backing of the gold substrate [40]. There are also two peaks labelled "impurity" that appear and grow in intensity over the course of the experiment at 949 and 1272 cm^{-1}. These peaks have previously been attributed as impurity peaks in the hydrolysis of UF_6 mentioned earlier [40]. In that study, the experimental apparatus contained Viton O-rings; once these O-rings were removed, these peaks disappeared in subsequent exposures. Therefore, it was concluded that UF_6 gas was attacking the Viton and producing these impurity peaks. The primary structure of Viton is a polymer of fluorocarbons. This means that UF_6 was attacking the polymer and cleaving it into smaller gas phase fluorocarbon chains. In our work, we see these peaks appear once UF_6 is introduced to the system and then grow over time of exposure. Our system is devoid of any Viton or similar substances. If this was just an impurity in the gas tank, we would expect the concentration of this peak to stay the same over time (as we see for the UF_6) or decrease as it adsorbed to the surface of the sample or the chamber walls. Because these peaks increase overtime, we believe that UF_6 is fluorinating the only source of hydrocarbon in the chamber, the SAM alkanethiolate chains. Since no CF_x modes are seen in the hydrocarbon experiments, the SAM chains must first be cleaved from the surface and enter the gas phase. UF_6 oxidizes and subsequently cleaves the sulfur head group at the Au-S bond, most likely at grain boundaries and defects in the surface. Once released from the substrate, the alkane chains could be fluorinated by the gas-phase hexafluoride. While we do not observe this at lower pressures of UF_6, this may be due to the low concentration in the system.

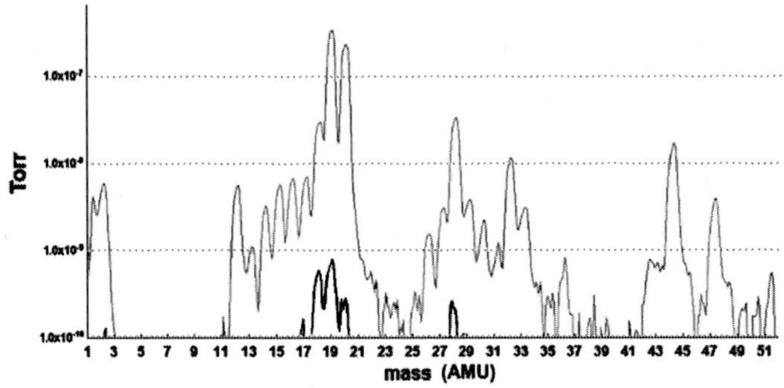

Figure 8. Residual gas analysis of the chamber evacuated to 10^{-8} Torr (dark trace) and the UF_6 tank headspace equilibrated with the UF_6 gas manifold (light trace).

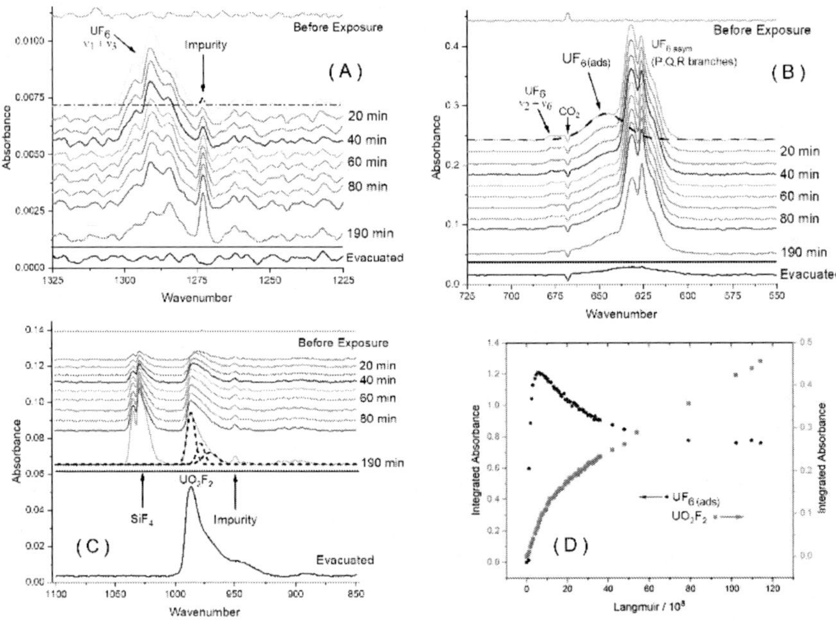

Figure 9. Detailed view of different regions (A) 1325-1225, (B) 725-550, and (C) 1100-850 cm^{-1} for the hydrocarbon SAM RAIRS spectra during 1.00 torr UF_6 gas exposure over time. Spectra labelled "Evacuated" are the resulting spectra after gas is removed and chamber pumped down., (D) Integrated absorbance of adsorbed UF_6 and UO_2F_2 for the hydrocarbon surface that was exposed to 1.00 torr UF_6 gas [42].

Now let's examine changes to the spectra that remain post-exposure. Two bands are seen for new surface-bound species – a broad band at ~640-630 cm^{-1} and a peak with shoulders at 1010-930 cm^{-1}. Seen in Figure 9B, the principle UF_6 ν_3 mode overlaps the shallow band centered at 647 cm^{-1}, which grows in during exposure, but is not fully removed upon evacuation. This peak is attributed to surface-perturbed ν_3 mode of adsorped UF_6 molecules. We executed quantum chemical electronic structure calculations (at the MP2/aug-cc-pVDZ-PP level) predict the shift in the IR spectra due to surface adsorption [43]. To achieve this, we did a full geometry optimization followed by normal mode frequency analysis of UF_6 complexing with a triad of methylamine molecules arranged to mimic the tips of SAM surface chains, as shown inset of Figure 10. We calculated the harmonic ν_3 stretching frequency for gas phase $UF_{6(g)}$ to be ~625cm^{-1}. Upon absorption to the SAM surface, our calculations predict a shift of about 40 cm^{-1} to about 665 cm^{-1}. UF_6 binds more strongly and is more distorted when interacting with the amine-terminated SAM than with the hydrocarbon SAM. As such, we would expect a less drastic, but still significant blue shift upon UF_6 adsorption ($UF_{6(ads)}$) to the surface. Tighter binding/stronger interactions of the $UF_{6(ads)}$ with the hydrocarbon SAM surface can be influenced as the pressure of $UF_{6(g)}$ is increased, as seen in Figure 10. Here we plot the maximum of the fitted peak for the $UF_{6(ads)}$ (see dashed line in Figure 9B) as a function of $UF_{6(g)}$ pressure. As can be seen, increasing $UF_{6(g)}$ from 0.01 to 1.0 Torr blue-shifts the $UF_{6(ads)}$peak position. Higher $UF_{6(g)}$ pressures may create a driving force for higher $UF_{6(ads)}$ coverage on the surface, resulting in more $UF_{6(ads)}$ molecules taking on geometries resembling our calculations. Only for the highest pressure of $UF_{6(g)}$ exposure does this $UF_{6(ads)}$ signal remain after evacuating the chamber – the physisorped UF_6 is removed when the low pressure UF_6 is pumped away.

Why though, does the weakly physisorped UF_6 remain attached to the surface after gas evacuation? One explanation may be that it becomes "trapped" during the formation of another surface-bound product we can monitor in the RAIRS spectra. Uranyl fluoride, UO_2F_2, is a water contaminant reaction product with respect to our goals here, but nevertheless forms on the surface of hydrocarbon and fluorocarbon SAMs at all but the lowest of pressures. Figure 9C shows the expected position for the U = O stretch in UO_2F_2 as an asymmetric peak between 1010-930 cm^{-1} [44-46]. UF_6 is known to react with water upon immediate contact and form UO_2F_2 and HF gas [40, 47-49], and we do see at the highest pressures evidence of some gas-phase HF formation in the 4250-3750 cm^{-1} range (not shown), but the signal is quite weak.

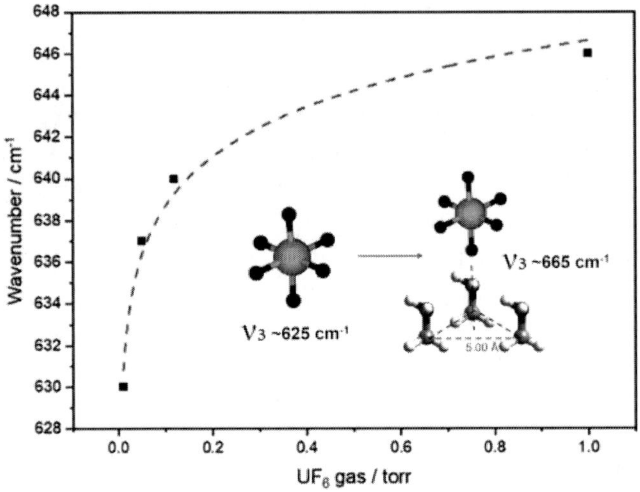

Figure 10. Peak position (in wavenumbers) of adsorbed UF_6 (see Fig. 9B) vs exposure of UF_6 gas (in torr). The inset illustrates the calculated vibrational mode of UF_6 upon adsorption to a simple surface model. [42].

Figure 9D shows that the integrated $UF_{6(ads)}$ intensity increases initially, but then decreases and plateaus as UO_2F_2 steadily continues to increase over time. Presumably the initially adsorbed UF_6 is partially converted to UO_2F_2 until this UO_2F_2 "protects" the remaining initial concentration of $UF_{6(ads)}$ from further oxidation. Perhaps this UF_6 is trapped deeper in the SAM chains. This initial increase in $UF_{6(ads)}$ was not as large on the fluorocarbon SAM, and did not show a subsequent decrease, but the signal there also plateaus as the UO_2F_2 signal continues to increase. The hydrocarbon surface seems to be able to initially accommodate more $UF_{6(ads)}$ than the fluorocarbon surface, but in both cases, $UF_{6(ads)}$ is limited and gives way to uranyl oxide formation.

3. Concluding Summary

The chemistry investigation reported herein was part of a larger effort to determine the feasibility of gas-phase UF_6 alpha spectroscopy for in-process enrichment measurements. The investigation had an applied component – the identification of a suitable passivating alpha detector coating material – and a basic science component – the elucidation of mechanisms of interaction between UF_6 and thin-film coating materials. This chapter has focused on the basic science component. It was determined that hydrocarbon and fluorocarbon SAM surfaces are mostly inert to reaction with WF_6 and UF_6 gas and that the relative adsorption strength of UF_6 on the different inert surfaces likely follows the trend Au > CH_3 > CF_3 > AlF_3. Furthermore, it was determined that water contamination in UF_6 gas causes deposition of UO_2F_2 on inert surfaces despite the extensive water removal efforts used in this work. UO_2F_2 mass deposition rates were primarily a function of surface roughness and reactant adsorption strength.

In terms of the applied component of the overall project, which we did not detail herein, AlF_3 was identified as an effective coating material due to its ability to protect the alpha detector surface while minimizing UO_2F_2 deposition [50]. A commercially available alpha detector was coated with a 3 nm layer of AlF_3 and UF_6 gas sample alpha spectroscopy experiments were performed. The coating was effective for protecting the alpha detector from damage in the UF_6 and produced HF gas and the detector continues to operate normally after several months in this environment. UO_2F_2 deposition on the detector surface due to water contaminants caused alpha particle attenuation that prevented the peak resolution required for enrichment measurements. Design recommendations were provided for a sacrificial chamber system to remove water from UF_6 samples. This method could also be useful for other UF_6 experimental systems, particularly systems intended for precision measurements where surface deposition of UO_2F_2 could affect the results. Testing of a new system with advanced moisture control features would be required to determine if sufficient water control could be achieved for gas-phase alpha spectroscopy enrichment measurements. For this reason, the feasibility determination of using gas-phase alpha spectroscopy for in-process enrichment measurements remains inconclusive.

4. Materials and Methods

4.1. Surface Preparation

SAMs used in this work were prepared by spontaneous chemisorption of the corresponding alkanethiol precursors from a 1 mM ethanolic solution onto cleaned gold. Gold substrates were either gold coated float glass (Dynasil Evaporated Metal Films, 100 nm Au with 5 nm Cr binder) or gold coated Quartz Crystal Microbalance (QCM) substrates (Phillip Technologies, Inficon

6 MHz AT Quartz). The substrates were cleaned in a piranha solution (70/30 (v/v) mixture of H_2SO_4/30% H_2O_2) prior to use. On these substrates we formed ω-functionalized SAMs. Hydrocarbon SAMs (CH_3-terminated SAMs) were created from either 1-octadecanethiol ($HS(CH_2)_{17}CH_3$) or 1-undecanethiol ($HS(CH_2)_{10}CH_3$); these SAMs are referred to as 18C and 11C CH_3 SAMs, respectively. A near-fully fluorinated SAM (CF_3-terminated SAM) was created from 1*H*,1*H*,2*H*,2*H*-perfluorodecanethiol ($HS(CH_2)_2(CF_2)_7CF_3$. An amine-terminated SAM (NH_2 terminated SAM) was constructed of 16-aminohexadecanethiol ($HS(CH_2)_{16}NH_2$), which is formed from the stable hydrochloride ($HS(CH_2)_{16}NH_3^+ Cl^-$) in solution. Gold surfaces were immersed in the alkanethiol solutions for at least 24 hours to ensure proper packing and ordering of the alkanethiol chains. Upon removal the surfaces were rinsed with ethanol and flushed with ultrahigh purity nitrogen. The amino SAMs were additionally rinsed with triethanolamine (TEA) to deprotonate the SAM and remove the chloride, and again rinsed with ethanol and dried under nitrogen. This TEA rinse ensures effective removal of the chloride anion, leaving the neutral amine group behind as confirmed with X-ray Photoelectron Spectroscopy (XPS) [51]. All precursor chemicals were purchased from Sigma Aldrich and used without further purification. New SAMs were periodically characterized with reflection-absorption infrared spectroscopy. A narrow absorption band centered at or below 2919 cm^{-1} for the asymmetric stretching mode of the methylene groups along the backbone of the alkane chains indicated that the SAMs were well-ordered and densely packed.

4.2. Hexafluoride Gas Exposure

UF_6 cylinders of depleted uranium, 3.159% LEU, and 10.418% LEU were acquired from New Brunswick Laboratory and used for this work. Tungsten hexafluoride (99.9%, Sigma Aldrich) was used in several experiments because it is a good surrogate for UF_6 in terms of chemical reactivity [10], and because samples exposed to WF_6 can be analyzed in non-radiological labs. WF_6 and UF_6 exposure experiments were carried out in the custom exposure and analysis system pictured in Figure 2 below, which has been described in more detail elsewhere [52]. The system includes two ultra-high vacuum (UHV) chambers; a sample loading chamber and a hexafluoride gas exposure chamber. The system has both internal infrared heating and external resistive heating to bake the chamber under vacuum to remove water and other contaminants from the chamber. The chamber is capable of reaching a minimum pressure of 3×10^{-9} Torr following bake-out.

Following evacuation, the chamber was analyzed by mass spectrometry using an on-instrument residual gas analyzer (RGA, range 1-64 amu). The system has a Fourier transform infrared (FTIR) spectrometer that can be used in transmission mode to study the partial pressures of UF_6, HF, and water. This spectrometer can also be used in reflection mode as described below. The exposure chamber houses a quartz crystal microbalance (QCM) for measuring deposition rates.

For the experiments reported here, samples were introduced through the sample loading chamber or directly into the exposure chamber. The chambers were evacuated to 10^{-8} Torr to remove water and other potential contaminates from the sample surface. Gold, AlF_3, or SAM coated surfaces were exposed to WF_6 or UF_6 at pressure ranging from 0.01 to 5.0 Torr for up to 35 days followed by evacuation of the chamber to UHV using an inert gas flushing procedure described previously [52]. When chamber pressure reached 10^{-6} Torr, the chamber components were analyzed by the RGA.

Figure 11. Hexafluoride gas exposure system with in-situ surface and gas analysis capabilities. In this image, the RAIRS beam path enclosure is removed to allow viewing the entire chamber. The beam is directed through an IR window that replaces the solid flange in the foreground [52].

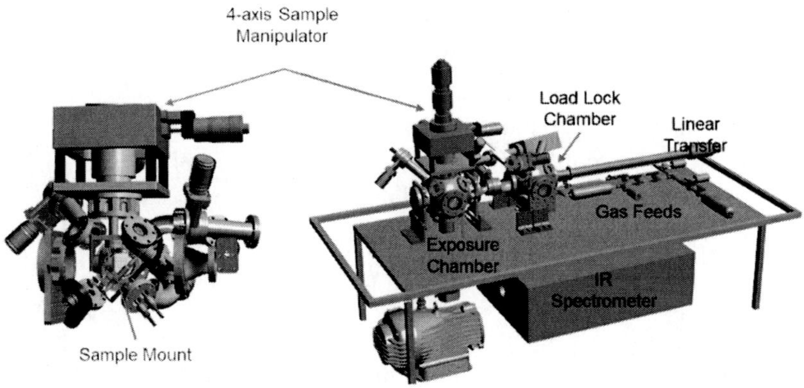

Figure 12. Multi-modal UF_6 materials research apparatus [52].

4.3. Quartz Crystal Microbalance

For the long-term QCM exposure studies, bare gold-coated QCM crystals, unmodified QCM crystals coated with 3 nm AlF_3, or SAM-modified gold-coated QCM crystals were inserted into the UF_6 exposure chamber drawer type QCM (CDS-A0F47, Inficon). The QCM was operated using an Inficon oscillator (783-500-013) and a Kurt J. Lesker Rate/Thickness Monitor (FTM2400-S2E). Before exposure, the chamber was evacuated to 10^{-8} Torr to remove surface contaminants. The various surfaces were exposed to UF_6 gas at 0.01, 0.1, or 1.0 Torr until a steady state frequency was reached. To achieve QCM measurements under a precisely controlled vacuum environment, no inert gas flush was used on the backside of the crystal. For this reason, the UF_6 data is reported as the negative frequency shift between UHV and steady state at each pressure.

For adsorption-desorption kinetics studies, samples were exposed to 5 second pulses of WF_6 gas in an atomic layer deposition (ALD) instrument, under a continuous purge of 1 Torr argon. The viscous flow ALD reactor is stainless-steel with a heated crossflow rectangular reactor with dimensions of 30 mm × 105 mm × 600 mm. The reactor flow tube was resistively heated externally to the desired bake-out or exposure temperatures. Unmodified Au, AlF_3 coated, or SAM-modified gold coated QCM crystals were loaded into a Maxtech QCM holder which is heated by radiation and convection from the hot walls. Ultrahigh purity argon was supplied via mass flow controllers with a maximum throughput of 200 sccm constantly flowing through the reactor which maintains background pressure of 1 Torr and purges the reverse side of the QCM. The reactor pressure is monitored by a heated 10 Torr Baratron capacitance manometer (MKS Instruments). To remove surface water the surfaces were baked under purge gas for 2 hours at 70°C for SAM surfaces or 250°C for the more robust Au and AlF_3 surfaces. WF_6 gas was supplied via computer controlled pneumatic valves with desire dose and purge time. QCM analysis of WF_6 exposure was performed at room temperature (28°C). Deposition was considered "irreversible" for the discussions herein if the deposited material was not removed during the inert gas flushing between hexafluoride gas pulses.

The surface densities of the monolayers were estimated by two methods. For the bare gold surface, the mass of a rigidly adsorbed WF_6 monolayer was estimated based on a molecular spacing of 7.7 Å (1.5 times the molecular diameter of WF_6) to give a total of 82.5 ng/cm^2. For the less rigid SAM-modified Au surfaces, the mass of a deposited monolayer was estimated based on the total mass of the non-volatile reaction products, assuming one surface reaction per self-assembled monolayer unit chain. The approximate coverage of the SAM unit chains on a gold surface was calculated to be 5.7×10^{14} chains per cm^2 [51, 53, 54] Surface densities of 263 ng/cm^2 and 280.9 ng/cm^2 were calculated for HF generating and non-HF generating reactions, respectively. It should be noted that the values only hold for chemically-bound or rigidly adsorbed molecules, so a monolayer of weakly adsorbed molecules would appear as a partial monolayer in the data.

4.4. Reflection-Absorption Infrared Spectroscopy

A Bruker Vertex 70 FTIR spectrometer operating in the mid-IR range (4,000 to 400 cm^{-1}) was used for the in-situ RAIRS studies. An external liquid nitrogen cooled MCT detector was used due to its high sensitivity for the low-signal reflection-absorption infrared spectroscopy signals. The infrared beam exited the spectral bench and was guided through the reaction chamber and to the detector by a custom series of mirrors made by Harrick Scientific Products, Inc. The beam exited the spectrometer with a diameter of one and half inches and traveled along mirrors before being directed into the chamber with a focusing mirror. The beam passes through a rotatable polarizer and enters the chamber through zinc selenide windows, where it is directed at a grazing angle of 85° to the surface normal of the sample. The sample was brought into the beam path by a 4-axis (x and y translation, rotation, and tilt) sample manipulator shown in Figure 12. As the beam was reflected off the sample, it exited the chamber and hit a collimating mirror, which directed the light into the detector. The RAIRS technique benefits from controlling the polarization since it is sensitive to dipole orientation. The polarization is easily switched between s-polarization, which is noted as being light perpendicular to the surface normal, and p-polarization, which is light parallel to the surface normal.

All mirrors and beam path tubes were purged with house nitrogen that had passed through a desiccating column. The detector house was kept dry by color-changing desiccant. When experiments were not being conducted, the reaction chamber was constantly pumped down to ultra-high vacuum by a turbomolecular pump. The reaction chamber pumps down to UHV before starting an experiment. Because UF_6 is reactive with water, a residual gas analyzer is used to monitor the partial pressure of water in the chamber. RAIRS spectra in this chapter are shown as difference spectra compared to an initial pre-exposure scan and are all taken with p-polarized light.

All spectra were the average of 128 scans at 2 cm^{-1} resolution, which takes about one minute to collect. Difference spectra were taken while a leak valve was opened and closed to introduce the hexafluoride gas to the desired pressure and were collected for several hours through the experiment to monitor changes to sample with respect to time. After the experiment, the chamber was purged several times with helium gas that was passed through a waste trap system to capture gaseous U products. The chamber was then pumped down again using the turbomolecular pump. Spectra were taken after the chamber was thoroughly evacuated to monitor any permanent changes.

Acknowledgments

The authors are thankful to Joseph Savina, Candido Pereira, Anil Mane, and Jeff Elam at Argonne National Lab, and Katherine A. Charbonnet and Felio Perez at The University of Memphis for their contributions in support of this work. This work was funded by the United States National Nuclear Security Administration Office of Defense Nuclear Nonproliferation Research and Development. Preparation of the manuscript was supported by a Summer Faculty Stimulus Grant from the College of Arts and Sciences at The University of Memphis. The High Performance Computing Center and the Computational Research on Materials Institute at The University of Memphis (CROMIUM) also provided generous resources for this research. Any opinions, findings, and conclusions or recommendations expressed in this book chapter are those of the author(s) and do not necessarily reflect the views of the funding/supporting agencies.

Further information about this work is contained in Kristin N. Knight's (now Kristin K. Anderson) PhD Dissertation, "Investigations of Uranium Hexafluoride Chemistry on Organic Surfaces" available at https://digitalcommons.memphis.edu/etd/3260.

Figures in this chapter have undergone graphical alterations/edits to conform to publication standards and are intended for use in this chapter as *illustrations*. The original referenced figures should be used for rigorous scientific examination/comparison.

This work would not have been possible without the intellectual contributions of Dr. Megan E. Bennett.

References

[1] Younkin J., Garner, J. *On-Line Enrichment Monitor Phase II Final Report,* ORNL/LTR-2017/108, 2017.

[2] Saniger, J., Alba, F., Garrido, S. J., Avendaño, E. The kinetics of aluminum-7075 corrosion by uranium hexafluoride, *Corrosion Science* 1990, *30*, 903-913.

[3] Morel, B., Selmi, A., Moch, L., Hiltbrunner, J.-M., Achour, M., Benzouaa, R., Bock, A., Jouffret, L., Bonnet, P., Hamwi, A., Dubois, M. Surface reactivity of uranium hexafluoride (UF_6), *Comptes Rendus Chimie* 2018, *21*, 782-790.

[4] Grosse, A. Chemical Properties of Uranium Hexafluoride, *Oak Ridge Technical Report,* 1941, NTIS Issue Number 201006.

[5] Bennett, M. R., L. M. Ferris, L. M. Formation of U(V) in a molten fluoride salt at 550–650°C by reaction of gaseous UF_6 with dissolved UF_4, *Journal of Inorganic and Nuclear Chemistry* 1974, *36*, 1285-1290.

[6] Eichler, R., Nikolay V Aksenov, Alexey V Belozerov et al., Thermochemical and physical properties of element 112, *Angewandte Chemie* 2008, *47*, 3262-3266.

[7] Launiere, C., Savina, J., Pereira, C., Alexander, W. A. Hexafluoride Chemical Compatibility Studies, Report for National Nuclear Security Administration Office of Defense Nuclear Nonproliferation Research and Development, ANL-19/07, Argonne National Laboratory, 2019.

[8] AREVA Eagle Rock Enrichment Facility Nuclear Regulatory Commission License Application, Safety Analysis Report Rev. 3, Ch. 6 Chemical Process Safety, 2011.

[9] Roy, O., Marquet, B., Alric, J. P., Jourdan, A., Morel, B., Langlois, B. R., Billard, T. (Re)Investigation of the reactivity of uranium hexafluoride toward several organic functions at room temperature, *Journal of Fluorine Chemistry* 2014, *167*, 74-78.

[10] Orekhov, V. Reaction of Organic Compounds with Fluorides of f- and d- Elements in Groups V and VI, *Russian Chemical Reviews* 1977, *46*, 799-827.

[11] Gromov, O. B., Khorozova, O. D., Travin, S. O., Atakhanova, E. L. Reactions of UF_6 and HF with Ethanol. *Radiochemistry* 2018, *60*, 152-157.

[12] Vergamini, P. J., Reaction of UF_6 with Methanol to Form UF_5OMe. *J. Chem. Soc., Chem. Commun.* 1979, 54-55.

[13] Olah, G. A., Shih, J. G., Prakash, G. K. S. Fluorine-Containing Reagents in Organic Synthesis. *J. Fluor. Chem.* 1986, *33*, 377–396.

[14] Atakhanova, E. L., Orekhov, V. T., Savel'ev, D. S., Khorozova, O. D., Khudin, A. S., Shiryaeva, V. V. Interaction of UF_6 and UF_4 with Methanol. *At Energy* 2016, *119*, 404-407.

[15] Wilson, P. W. Reactions and Comparative Reactivity of Uranium Hexafluoride. *Rev. Pure Appl. Chem*. 1972, *22*, 1-12.

[16] Olah, G. A., Welch, J., Ho, T. L. Synthetic Methods and Reactions. 17. Uranium Hexafluoride, a Convenient New Oxidizing Agent for Organic Synthesis. *J. Am. Chem. Soc*. 1976, *98*, 6717-6718.

[17] Olah, G. A., Welch, J. Synthetic Methods and Reactions. 46. Oxidation of Organic Compounds with Uranium Hexafluoride in Haloalkane Solutions. *J. Am. Chem. Soc.* 1978, *100*, 5396-5402.

[18] Olah, G. A. Some New Synthetic Reagents and Reactions. *Acc. Chem. Res.* 1980, *13*, 330-336.

[19] Goosen, A., McCleland, C. W., Venter, P. J., Venter, M. W. Reactions of Uranium Hexafluoride with Organic Substrates. *S. Afr. J. Chem.* 1987, *40*, 30-34.

[20] Schnautz, N. G., Venter, P. J. Uranium Fluoride Chemistry. Part 1. The Gas Phase Reaction of Uranium Hexafluoride with Alcohols. *S. Afr. J. Chem.* 1992, *45*, 59-62.

[21] O'Donnell, T. A., Stewart, D. F., Wilson, P. Reactivity of Transition Metal Fluorides. II. Uranium Hexafluoride. *Inorg. Chem.* 1966, *5*, 1438-1441.

[22] Olah, G. A., Welch, J. Synthetic Methods and Reactions; XXV. Tungsten Hexafluoride, a Mild Oxidizing Agent for the Cleavage of Hydrazones. *Synthesis* 1976, 809-810.

[23] Zhang, Y., Chapleski, R. C., Lu, J. W., Rockhold, T. H., Troya, D., Morris, J. R. Gas Surface Reactions of Nitrate Radicals with Vinyl-Terminated Self-Assembled Monolayers. *Phys. Chem. Chem. Phys*. 2014, *16*, 16659–16670.

[24] Lu, J. W., Day, B. S., Fiegland, L. R., Davis, E. D., Alexander, W. A., Troya, D., Morris, J. R. Interfacial Energy Exchange and Reaction Dynamics in Collisions of Gases on Model Organic Surfaces. *Prog. Surf. Sci*. 2012, *87*, 221–252.

[25] Fiegland, L. R., McCorn, M., Fleur, S., Morris, J. R. *Reactions of C d C-Terminated SelfAssembled Monolayers with Gas-Phase Ozone*. 2005, No. 9, 2660–2661.

[26] Alexander, W. A. Particle Beam Scattering from the Vacuum-Liquid Interface. In *Physical Chemistry of Gas-Liquid Interfaces*; Faust, J. A., House, J. E., Eds., Elsevier: Amsterdam, The Netherlands, 2008.

[27] Rapp, K. Pattern of explosive reaction between uranium hexafluoride and hydrocarbon oils, *Oak Ridge Gaseous Diffusion Plant*, K/GD--1631-REV1, 1986.

[28] Charbonnet, K. A., Alexander, W. A. Computational Studies of Uranium Hexafluoride Interacting with Functionalized Organics. I. Screening Intermolecular

Potentials between UF_6 and Small Organic Functional Groups. *J. Phys. Chem. A* 2021, *125*, 10130-10137.

[29] Sherrow, S. A., Hunt, R. D. FTIR Spectra of the Atmospheric Hydrolysis of Uranium Hexafluoride. *Appl. Spectrosc.* 1991, *45*, 1008-1016.

[30] Alexander, W. A., Troya, D. Theoretical Study of the Dynamics of Collisions Between HCl and ω-Hydroxylated Alkanethiol Self-Assembled Monolayers. *J. Phys. Chem. C* 2011, *115*, 2273- 2283.

[31] Fukushima, H., Seki, S., Nishikawa, T., Takiguchi, H., Tamada, K., Abe, K., Colorado, R., Graupe, M., Shmakova, O. E., Lee, T. R. Microstructure, Wettability, and Thermal Stability of Semifluorinated Self-Assembled Monolayers (SAMs) on Gold. *J. Phys. Chem. B* 2000, *104*, 7417–7423.

[32] Lenk, T. J., Hallmark, V. M., Hoffmann, C. L., Rabolt, J. F., Castner, D. G., Erdelen, C., Ringsdorf, H. Structural Investigation of Molecular Organization in Self-Assembled Monolayers of a Semifluorinated Amidethiol. *Langmuir* 1994, *10*, 4610–4617.

[33] Tamada, K., Knoll, W., Colorado, R., Graupe, M., Shmakova, O. E., Lee, T. R. Molecular Packing of Semifluorinated Alkanethiol Self-Assembled Monolayers on Gold: Influence of Alkyl Spacer Length. *Langmuir* 2001, *17*, 1913–1921.

[34] Day, B. S., Morris, J. R., Alexander, W. A., Troya, D. Theoretical study of the effect of surface density on the dynamics of Ar+ alkanethiolate self-assembled monolayers. *J. Phys. Chem. A* 2006, *110,* 1319-1326.

[35] Peruski, K. M., Davis, T. A., Chan, G. C. Y., Mao, X., Trowbridge, L., Martin, L. R. Cell design for laser-induced breakdown spectroscopy measurements on reactive gas samples, *Journal of Fluorine Chemistry* 2022, *255–256*, 109951.

[36] Milin, M., Rannou, J., Viaulle, L., Caplin, G., Evo, S. "Hydration of uranium residues contained in enriched UF_6 cylinders," *Proceedings of the 18th International Symposium on the Packaging and Transportation of Radioactive Materials PATRAM Kobe*, Japan, September 18-23, 2016.

[37] Kessie, R. Plutonium and Uranium Hexafluoride Hydrolysis Kinetics, *Ind. Eng. Chem. Process Des. Dev*. 1967, *6*, 105–111.

[38] Hu, S., Lin, H., Wang, X.-Y., Chu, T.-W. Effect of H_2O on the hydrolysis of UF_6 in the gas phase, *J. Mol. Struc*. 2014, *1062*, 2014.

[39] Sherrow, S., Hunt, R. FTIR spectra of the hydrolysis of uranium hexafluoride, *J. Phys. Chem.* 1992, *96*, 1095-1099.

[40] McDowell, R. S., Asprey, L. B., Paine, R. T. Vibrational Spectrum and Force Field of Uranium Hexafluoride. *J. Chem. Phys*. 1974, *61*, 3571–3580.

[41] Knight, K. N. "Investigations of Uranium Hexafluoride Chemistry on Organic Surfaces" Dissertation, The University of Memphis, 2022. *Electronic Theses and Dissertations*. 3260. https://digitalcommons.memphis.edu/etd/3260.

[42] Frisch, M. J., Trucks, G. W., Schlegel, H. B., Scuseria, G. E., Robb, M. A., Cheeseman, J. R., Scalmani, G., Barone, V., Petersson, G. A., Nakatsuji, et al. Gaussian 16, *Revision B.01,* Gaussian, Inc., Wallingford CT, 2016.

[43] Marshell, J. Infrared Investigation of Uranyl Chloride and Uranyl Fluoride. *Asian J. Chem*. 2007, *19*, 4645–4648.

[44] Kirkegaard, M. C., Langford, J., Steill, J., Anderson, B., Miskowiec, A. Vibrational Properties of Anhydrous and Partially Hydrated Uranyl Fluoride. *J. Chem. Phys.* 2017, *146*, 024502.

[45] Hoekstra, H. R. Uranium–Oxygen Bond Lengths in Uranyl Salts: Uranyl Fluoride and Uranyl Carbonate. *Inorg. Chem.* 1963, *2*, 492-495.

[46] Armstrong, D. P., Bostick, W. D., Fletcher, W. H. An FT-IR Study of the Atmospheric Hydrolysis of Uranium Hexafluoride. *Appl. Spectrosc.* 1991, *45*, 1008–1016.

[47] Reiner, R. H., *Moist Air Reaction with Excess* UF_6. 1989. United States. https://www.osti.gov/servlets/purl/10103576.

[48] Richards, J. M., Martin, L. R., Fugate, G. A., Cheng, M. D. Kinetic Investigation of the Hydrolysis of Uranium Hexafluoride Gas. *RSC Adv.* 2020, *10*, 34729–34731.

[49] Launiere, C. A, "Near Real-time Determination of UF_6 Enrichment Levels Final Report," *Report for National Nuclear Security Administration Office of Defense Nuclear Nonproliferation Research and Development*, ANL-20/42, Argonne National Laboratory, 2020.

[50] Bennett, M. E., Alexander, W. A., Lu, J. W., Troya, D., Morris, J. R. Collisions of Polar and Nonpolar Gases with Hydrogen Bonding and Hydrocarbon Self-Assembled Monolayers. *J. Phys. Chem. C* 2008, *112*, 17272-17280.

[51] Launiere, C., Savina, J., Pereira, C., Alexander, W. A. Construction of a Working UF_6 Exposure Chamber, *Report for National Nuclear Security Administration Office of Defense Nuclear Nonproliferation Research and Development,* ANL-18/32, Argonne National Laboratory, 2018.

[52] Hinterwirth, H. Quantifying Thiol Ligand Density of Self-Assembled Monolayers on Gold Nanoparticles by Inductively Coupled Plasma–Mass Spectrometry, *ACS Nano* 2013, *7*, 1129-1136.

[53] Day, B. S., Morris, J. R., Alexander, W. A., Troya, D. Theoretical study of the effect of surface density on the dynamics of Ar+ alkanethiolate self-assembled monolayers. *J. Phys. Chem. A* 2006, *110,* 1319-1326.

Chapter 4

Intermolecular Binding Potentials and Lewis Acid/Base Interactions of Uranium Hexafluoride - Organic Molecule Complexes

**William A. Alexander*
and Katherine A. Charbonnet**
Department of Chemistry, The University of Memphis, Memphis, TN, USA

Abstract

The stockpiles of depleted uranium hexafluoride waste may be useful for some as yet un-realized chemistry. Investigation of the intermolecular interactions between UF_6 and organic functional groups may glean trends and insight to inspire researchers to explore some of these potential chemistries. In this chapter, we discuss intermolecular binding energies (IBEs) predicted by quantum chemical calculations for UF_6 interacting with various functionalized small organic molecules. Geometries were derived from Moller-Plesset theory, with overall binding energies calculated with dispersion-corrected Density Functional theory methods. The interactive behavior is characterized based on the functional group constituents and is found to be highly contingent on a combination of conformational deformation, stabilizing non-bonding interactions, and sterics. The most important factor seems to be access to lone pair electrons in the organic functional groups, that interact strongly with the uranium UF_6 core. From these calculations, we also examine the potential of UF_6 as a Lewis acidity probe, which may further inspire synthetic chemistry schemes.

* Corresponding Author's Email: w.alexander@memphis.edu

In: The Chemistry of Uranium
Editor: William A. Alexander
ISBN: 979-8-89530-802-8

Keywords: uranium hexafluoride, intermolecular potentials, Lewis acidity, organic functional groups, density functional theory

1. Introduction

The interactions of UF_6 with various functional groups may be of wide interest. Uranium enrichment activities have generated large stores of depleted UF_6 "waste" that remains largely unused. Approximately 800,000 metric tons of depleted UF_6 is currently stored and undergoing conversion for eventual disposal [1]. Estimates for completion of this conversion and disposal process could exceed 25 years [1]. In the intervening time, exploring potential applications for this vast UF_6 resource could be valuable.

Computational study of UF_6 interactions with a wide range of compounds and materials may help reveal potential uses for this untapped resource. While we are thusfar unaware of any previous work examining the interaction of UF_6 with small organics, some work has examined UF_6 interactions with small non-organic gases. One of the earliest studies of UF_6 dimers was in 1974 by Schneider et al., who used statistical models to describe the UF_6-UF_6 and UF_6-Ng (Ng = Ar,Kr) interactions, with comparison to experimental observations [3]. Beu and co-workers have studied the clusters of $(UF_6)_n$ and the $(UF_6)_2$ dimer within Ar_n clusters for spectroscopy applications [4]. The UF_6 dimer has continued to be a topic revisited by multiple groups, both experimentally and computationally [5-7]. One potential reason depleted UF_6 remains unused is that the gas-phase UF_6 reacts with available water vapor to form uranium oxide solids. As such, the important UF_6-H_2O interaction has been studied in some detail [8-11].

In this chapter, we expand the scope of investigated UF_6 interactions by computationally screening geometry-optimized dimers of over 40 different functionalized organic small molecules. This computational screening approach allows us to evaluate the potential of various functional groups to interact with UF_6. The calculated intermolecular binding energies of UF_6–functional group "dimers" can be extrapolated with reasonable confidence to predict general trends for larger molecules and materials. This approach is a safe and cost-effective means of identifying promising candidates for experimental investigation. The validity of using such simplified molecular representations to model gas–organic surface interactions in has been demonstrated in previous studies [12-21].

These dimer geometries are analyzed with respect to deviations from the optimized structures of the lone molecular partners. Results are discussed in terms of the functional group interactions and conformation, especially with respect to accessible lone electron pairs, among other energetic aspects. Additionally, the Lewis acidity behavior is examined in correlation with the BF_3 affinity scale. These data should allow for informed selection of functional groups for experiments and analyses where UF_6 may be interacting with various surfaces, materials, or gases.

2. Computational Methods

Details of the computational methodology were reported previously [22]. Briefly, a "dimer" between UF_6 and a small organic molecule bearing a specific functional group molecule (FG) was geometry-optimized using MP2 theory and correlation-consistent basis sets of double-zeta quality, augmented with diffuse functions, and employing core-level pseudo potentials where appropriate.

Figure 1. Aerial view of the Portsmouth, Ohio, USA Gaseous UF_6 Diffusion Plant UF_6 cylinder storage area in 1991 [2]. The site is now being used as a depleted UF_6 conversion facility.

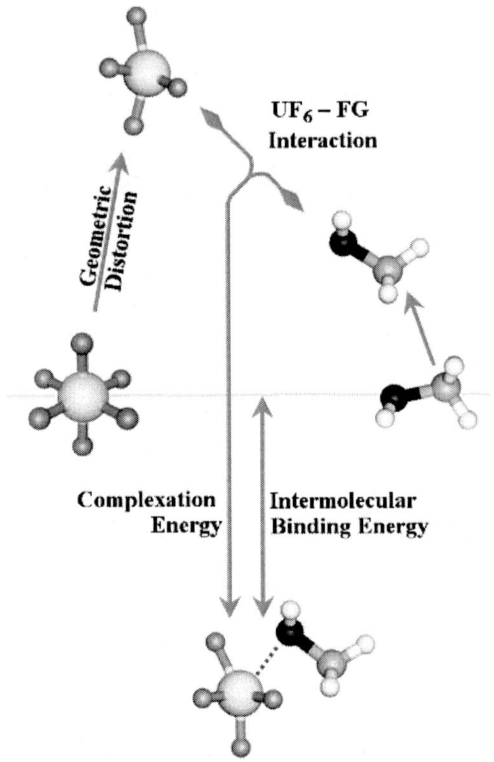

Figure 2. Schematic of the energy profile for UF_6 and a functional group molecule forming a dimer complex.

Table 1. Calculated complexation energies and IBE values for selected UF_6-FG pairs at the DFT/DZ level of theory[a]

Functional group	Complexation	IBE
CH_3CH_3	1.69	1.23
CH_3NH_2	25.66	14.47
CH_3OH	16.90	6.60
CF_3CF_3	1.20	1.08

[a]Reported energies in kcal/mol. The reported values are the "corrected" via counterpoise calculations on the dimer species. The IBE values are those discussed elsewhere in the text. DZ basis set: aug-cc-pwCVTZ (non-U atoms) and cc-pVTZ-PP (U).

For each FG-UF_6 pair, we obtained electronic energies of the FG and UF_6 alone, independently optimized, and for the optimized UF_6-FG dimer. Intermolecular Bond Energies (IBEs) were calculated by taking the difference in computed heats of formation between the separated, independently optimized UF_6 and FG molecules and the corresponding UF_6-FG dimer, as illustrated in Figure 2. To keep a concise and consistent definition throughout the chapter, we refer to and discuss IBEs that derive from density functional theory using double-zeta basis sets. The choice of this method and basis is based on comparison between results with the different methods and basis sets. MP2 and DFT calculations were performed with the double and triple zeta basis sets [23-27] for benchmarking comparisons. Benchmarking of different computational schemes revealed that the basis set superposition error must be corrected to obtain accurate IBE values [28]. However, vanishing returns were seen when moving from a double- to triple-zeta level of basis set. Therefore all IBEs are presented here using a DFT/DZ level of theory. Specifically we employed counterpoise-corrected calculations at the B3LYP level [29], with the G3BDJ model for dispersion corrections and damping [30], as implemented in Gaussian [31]. Further details can be found elsewhere [22].

It is important to note that for any strongly-interacting UF_6-FG pair, a large energetic barrier due to geometric distortion is overcome in the complexation of the two molecular partners, as seen schematically in Figure 1. Very different values are obtained if one takes the energy difference between the species without optimizing them separately. The difference between what we term the "complexation energy" and the "intermolecular bond energy" is given in Table 1 for ethane, ethylamine, methanol, and perfluoroethane molecules. Henceforth in the chapter we will only tabulate IBE values.

3. Results and Discussion of UF_6-Organic Molecule Interactions

The structure of gas-phase UF_6 has been extensively studied computationally [32-33]; UF_6 exhibits O_h symmetry with a bond distance of approximately 2 Å [34]. Peterson's high-accuracy CCSD(T) calculation resulted in a U-F bond distance of 1.999 Å using his cc-pVQZ-PP basis set for U and aug-cc-pVQZ basis set for F [25]. By trying combinations of smaller basis sets, the calculated MP2(FC1) U-F bond distance of 2.005 Å came closest to literature values when using the cc-pVDZ-PP and aug-cc-pwCVDZ basis sets for U and F. Structural investigations using electron diffraction techniques and high-resolution IR spectroscopy show UF_6 as having bond distances of 1.999 ± 0.003 Å, 1.996 ± 0.008 Å, and 1.9962 ± 0.0007 Å [34].

3.1. Overall Trends in UF_6-FG Dimers

We have obtained and report here the intermolecular binding energies (IBEs) of over UF_6-FG (functional group) dimers optimized with double-zeta quality DFT calculations. These results are tabulated for reference in Table 2. Further details of these calculations are available in [22] and supporting materials there. To facilitate summary of overall trends, we categorize species based on the heteroatomic substituents of the FG molecules and discuss average values and behaviors for these categories; clearly this analysis can be biased by the group size and specific molecule selection, but it is illustrative of the overall general behavior across the molecular classes.

Hydrocarbons display weak intermolecular binding energies (IBEs) with UF_6 (average of 2.3 kcal/mol), but the haloalkanes are the weakest class (average of 1.8 kcal/mol). Including an oxygen substituent increases the UF_6-FG intermolecular attraction by more than double (average of 5.5 kcal/mol) and including a nitrogen substituent further increases the interaction (average of 8.6 kcal/mol). Interestingly, including both O and N substituents results in values (average of 6.39 kcal/mol) similar to those with only O. Examining the minimum or maximum values (and hence the range) within the categories also follows this general trend.

Inspection of individual optimized dimer geometries indicates that U has a high affinity to interact with lone pairs, and it appears that the nature of the single lone pair on N can interact more easily than the lone pairs on O atoms. It is likely that the single N lone pair is able to interact with UF_6 in geometries that are better able to maintain symmetry elements than with O atoms, with the symmetric structures allowing other atomic substituents to more equitably participate in further additive non-covalent attractions between the dimer molecules.

Table 2. Calculated IBE values for all functional groups[a]

Functional group	IBE	Functional group	IBE
methane	1.23	methyl-isocyanate	2.35
ethane	1.67	ethyl-isocyanate	2.78
ethene	2.31	methyl-cyanate	3.85
ethyne	2.08	ethyl-cyanate	4.38
benzene	3.28	nitroso-methane	3.12
imidazole	13.49	nitroso-ethane	3.69
N-methyl imidazole	14.94	methanal	3.39
methyl amine	14.47	ethanal	4.27
ethyl amine	14.85	dimethyl-ether	6.28
ethane-nitrile	3.65	methanol	6.60
propane-nitrile	4.06	ethanol	5.86
methyl-isonitrile	4.11	ethanoic acid	5.11
ethyl-isonitrile	4.59	propanone	6.20
propan-2-one oxime	9.09	methyl-ethanoate	6.31
E-butan-2-one oxime	10.09	methanethiol	5.11
1*Z*-ethanal-oxime	3.92	fluoroethane	1.68
Z-propanal-oxime	4.16	perfluoro-methane	0.76
ethan-amide	9.24	hexafluoro-ethane	1.21
propan-amide	10.33	chloro-methane	2.65
azido-methane	3.34	bromo-methane	2.89
N'-hydroxy-ethanimidamide	3.19	*Z*-N'-hydroxy-propionimidamide	4.13
N-methyl-acetamide	10.60	N,N-dimethyl acetamide	12.34
N-methyl-propanamide	9.02	N,N-dimethyl propanamide	8.78

[a]Intermolecular Binding Energies given in kcal/mol. The values are the counterpoise-corrected DFT/DZ data as described in the text.

3.2. Distortion of UF_6 Geometries in FG Dimers

When looking at rendered images of geometries of UF_6 paired with each functional group, the distortion of UF_6's optimized octahedral geometry becomes evident, particularly when it comes to groups that resulted in deep potential energy wells. The fluorine atoms bend away from the approaching functional group in order to accommodate the shifting of uranium's electron density to complex with the incoming molecule. This geometrical distortion is greatest in highly polar functional groups with lone pairs, and much less obvious for alkyl and halogen groups. One way to characterize this is with the three F-U-F "anti" bond angles that would be 180° in the optimized lone UF_6 molecule. In Figure 3, we list these three angles for a few weakly and strongly distorting UF_6-FG pairs. Rather than list all three of these angles throughout the chapter, we use a short-hand, and discuss the sum angular deviation (SAD) of the molecule. The angular deviations are found by evaluating all possible F-U-F angles (n=15) for their deviation from the expected angle in a lone UF_6 molecule (90° or 180°). The sum angular deviation is simply the sum of all these deviations (i.e., the difference between the sum of all 16 F-U-F angles, subtracted from the expected 3×180° + 12×90° = 1620°). We therefore can speak about the UF_6 molecule in one complex being more or less SAD than in another.

The SAD value for methane (Figure 3a) is 3.10, inducing very little deviation in the UF_6 geometry upon interaction in the dimer complex. This is contrasted with the methylamine (Figure 3b) example, which has a SAD value of 43.96. These numbers confirm quantitatively what visual inspection makes obvious, illustrating the utility of SAD values as a proxy measure for geometric distortion. In general, as the UF_6 becomes more distorted, it allows the FG pair to make a closer approach to the U atom. For example, this is borne out in the r_{dist} between the U and the nearest non-H atom for the species in Figure 3; methane has an r_{dist} value of 3.95 Å, while the N atom on methylamine is almost 1.5 Å closer at a distance of 2.55 Å.

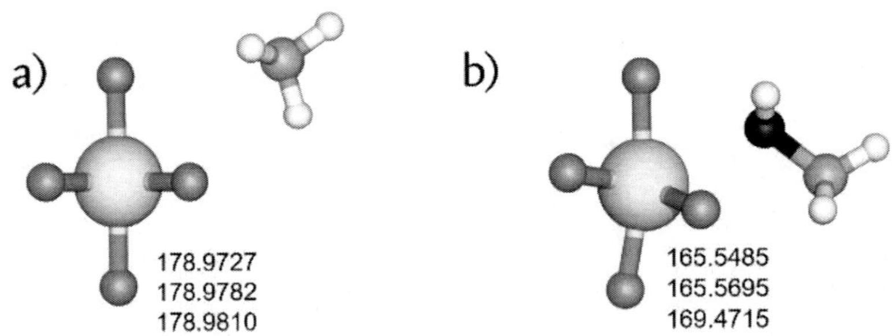

Figure 3. UF_6-FG dimers illustrating small (a) and large (b) UF_6 geometric distortion. The structures shown are UF_6 dimers with (a) methane and (b) methylamine. Inset values are the three angles in degrees found between F atoms on opposing sides of the U atom, listed in ascending order; in the lone optimized UF_6 molecules all three would be 180°. Adapted with permission from [22].

3.3. Hydrocarbon Chain Length Effects in UF_6/FG Pairs

Methane, the smallest hydrocarbon, shows one of the weakest interactions with UF_6 of all the possible species examined in this study, with an IBE of 1.23 kcal/mol. The addition of another methyl group going from methane to ethane contributes a further ~0.45 kcal/mol of stabilization energy for the UF_6 – FG dimer. These data indicate that UF_6 interactions are dominated by only weak, non-covalent vdW interactions for hydrocarbons.

It follows that increasing the size of the hydrocarbon backbone for other FG species should show a small increase in IBE values from the increased vdW effects. Represented within the small molecules we have complexed with UF_6, there 17 different pairs that vary by one methyl group (e.g., methyl cyanate vs. ethyl cyanate). On average, over these 17 pairs, the addition of the "ethyl methylene" group increases the IBE by ~0.34 kcal/mol. This is smaller than the ~approx. 1–2 kcal/mol contribution to the dispersion interaction per methylene unit in closely packed hydrocarbon layers [14, 17, 20]. Inspection of the various UF_6-FG dimers reveals that the additional methylene group is generally oriented away from the UF_6 molecule when the "added methylene group effect" is lower, and the additional methylene is oriented close to the UF_6 molecule when the effect is strong. A specific example is with the case of methanal vs. ethanal – addition of the 2nd methylene group enables the additional methyl group to interact well with the UF_6, raising the IBE by ~0.88 kcal/mol, which is significantly higher than the 0.34 kcal/mol average for the entire test set. This increase is because of the ability of the 2nd methyl group to interact closely with the UF_6, increasing vdW interactions.

Overall, the changes in IBEs (ΔIBE) upon adding a methyl group seem to stem from the relative size of the FG molecule and/or the influence of electron-donating character of the extra methylene group being able to contribute to an electron-poor FG. IBE values increase 25-35% when going from methane to ethane, or from methanal to ethanal, which should not be surprising since adding "more molecule" will add more vdW interactions. Because the molecule remains small, each atomic substituent is still close enough to UF_6 to significantly contribute to the stabilizing IBE. However, the stabilizing effect of the additional hydrocarbon component continues whether the added moiety is co-located to directly interact with UF_6 or not.

To see this effect, in other work we optimized propyl, butyl, pentyl, and hexyl amine to compare to methyl and ethyl amine. The longer amines all have their alkyl chains in the all-trans configuration pointing anti to the UF_6-amine intermolecular direction. IBEs increase from methyl to hexyl amine from about 14.5 to 15.3 kcal/mol. The addition of the ethyl methylene contributes 0.37 kcal/mol to stabilizing the dimer, the propyl contributes an additional 0.25 kcal/mol, and subsequent addition of methylene units contribute about 0.1 kcal/mol each. We hypothesize that at some length the chain-like molecule might like to "wrap-around" the UF_6 to increase interactions, and it is possible the straight-chain all-trans geometry is merely a local minimum. However, for the still-short hexyl amine, the required cost of a gauche defect is on the order of 3 kcal/mol, which is singularly larger than the difference in methyl/hexylamine IBEs and likely negates this notion.

3.4. UF_6 Forms Weak Covalent Interactions with FGs Bearing Only Nitrogen Heteroatoms

Species containing nitrogen form pairs with the most attractive potentials across the board, with imidazole and amine functional groups having the largest IBEs of over 14 kcal/mol. The other N-only functional group types all show IBEs of 4-5 kcal/mol, with the weakest IBEs in the class still being at or above the strongest investigated hydrocarbon IBEs. The IBE values roughly correspond with the increases in SAD values for the molecules. The ethane nitrile complex has a SAD value of 43.6, which increases for the amine and imidazole to 49.6 and 44.0, respectively. Addition of a methyl group increases the IBE and SAD values marginally for all of these species. The amine and imidazole UF_6 complexes have the strongest IBE values of all the FGs we studied, but there are a few complexes that display higher SAD values. We can contrast these strong IBE nitrogen complexes with regular alkanes, which display among the weakest IBEs.

As mentioned, the UF_6 distorts to accommodate the incoming FG molecule generally as the IBE increases. But why do the nitrogen moieties distort the UF_6, and the alkanes do not? Visualizing the molecular orbitals helps us infer the electronic distribution in the UF_6-FG pairs. In Figure 4, we have plotted the HOMO orbital for the UF_6-ethane and UF_6-methyl amine pairs for comparison. Two isosurfaces are plotted at the same two isovalues for each complex. As is seen in panel (a), the ethane molecule shares its most labile valence electrons very minimally with the three F atoms that form a facial plane toward the ethane. The vast majority of the electronic density stays localized on the ethane molecule, confirming that this interaction is clearly a non-bonding one. In contrast, we see in Figure 4(b) that a quite large density has shifted from the methyl amine to the UF_6 molecule, with all F atoms receiving what appears to be the same or more density than in the ethane case. Unlike with ethane, density also shifts to the uranium atom, with what appears as a node at the U center and density concentrating along the intermolecular axis. The nodal plane located across the methyl amine bond also becomes more pronounced and fully non-contiguous in comparison to the ethane. The lone pair on the nitrogen is sharing its density throughout the UF_6 atom, forming weak covalent interactions beyond the vdW interactions seen in the ethane molecule. (Recall that even though these are weak covalent interactions, covalent bonding is itself much stronger – circa an order of magnitude or more – than simple non-covalent vdW interactions.) It is clear now from the MOs that the UF_6 distorts consistent with VSEPR theory. As the nitrogen-containing functional groups begin to form weak ligand interactions with the uranium atom, the U-F bonds are repulsed, and increase their SAD depression from the optimal geometry.

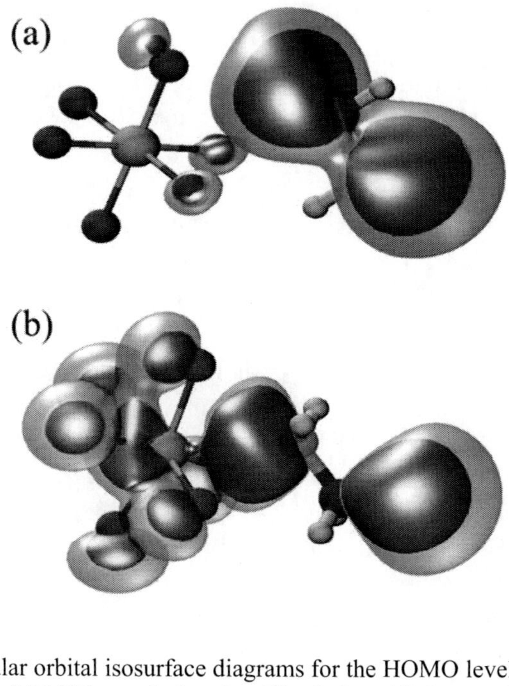

Figure 4. Molecular orbital isosurface diagrams for the HOMO level experienced by (a) UF_6-ethane, and (b) UF_6- methyl amine. The dark and light gray isosurfaces correspond to isovalues of -0.015 and -0.005, respectively. In this figure, the relative atom and bond sizing for the UF_6 is intentionally smaller than in other figures to better allow the MO surfaces to be seen. Adapted with permission from [22].

3.5. UF_6 Interactions with Oxygenated FGs Depend on Their Oxidation State

Introducing oxygen-only functionalities into the UF_6-FG dimer is not as effective at creating strong IBEs as nitrogenic species. The strongest IBE of this group was found to be only 6.60 kcal/mol for methanol, despite its relatively large SAD value 43.30. All the oxygenated FG molecules displayed IBEs that were correlative with the overall size/mass of the FG, in categories corresponding to whether a hydroxyl, carboxyl, or ether group was the moiety of closest approach to the UF_6. Using these categories, a plot of IBE value vs. mass (not shown) indicates a slope of approximately 0.05. By extrapolating back to the oxygen mass of 16 with this slope, we arrive at a base IBE value for O acting as an ether, hydroxy, or carboxyl of 5.0, 4.8, 3.1 kcal/mol, respectively. Subsequent mass addition to the molecule (increasing vdW interactions) increases the IBE values approximately linearly from these base values. The alkyl chain-length analysis indicates that this trend should likely attenuate once the molecule is around a "butyl" size, but the overall idea serves to reinforce the fact that the IBE is primarily dependent on the oxidation state of the oxygen in the FG, and not the "extra" atoms in the small molecule.

3.6. Species Containing Both Nitrogen and Oxygen as Heteroatoms Cause Largest UF_6 Distortions, Which Can Be Appreciated with Energy Diagrams

In this class species that contain both oxygen and nitrogen groups, the substituted amides and branched oximes display the largest IBE values, in the range of ~9-12 kcal/mol, approaching the strongest IBEs seen in the nitrogen-only class. N,N-dimethyl propanamide also has the largest SAD values of all species studied at 62.6; the singularly-substituted N-methyl propanamide is the second-most distorted with SAD values of 58.8. As these are the maximum deviations seen from the optimal UF_6 geometries, it is instructive to consider here the strain that UF_6 would otherwise feel if not complexed with the FG molecules. Comparing single point calculations (at the same DFT level of theory) of the skewed UF_6 geometry seen in the N,N-dimethyl propanamide complex to its solo optimized geometry indicates that UF_6 would experience a destabilizing energy of +19.95 kcal/mol. This corresponds to more than twice the magnitude of the UF_6–FG IBE of (-)8.27 kcal/mol, but in the opposite direction! The same analysis for the N,N-dimethyl propanamide molecule reveals that the conformational change in the FG only experiences +2.53 kcal/mol destabilization from its optimal lone geometry. This concept is easily visualized when built into an energy diagram; such plots are found in Figure 5 for simple UF_6-FG pairs representing a variety of IBE strengths and functional group class. The propanamide's hefty destabilizing distortion upon complexation with UF_6 can be further appreciated in comparison with these other species. Ethylamine (Figure 5a) and methyl imidazole (not shown) induce UF_6 upon complexation to take on conformations destabilized by 11.17 and 12.51 kcal/mol relative to their optimized geometry, all while the FGs remain in a nearly thermoneutral conformation. In cases where the FG dimer results in any significant IBE values, we find the UF_6 quite distorted, while the FG itself is rarely significantly disturbed. We can contrast this to behavior of the ethane (Figure 5b) and perfluoromethane (Figure 5d) dimers; these species have very small IBE values and minimal SAD values, so it is not surprising that they have nearly thermoneutral energy diagrams. Another indicator of strong binding interactions is the atomic partial charge distribution found in the molecular pairings. We use as a measure of this the Mulliken charge analysis values for the partial charges summed for the UF_6 and FG fragments. The largest charge transfer in our set was seen in the dimethyl propanamide, where 0.42 e^- is transferred from the FG to the UF_6. In comparison, the charge transfer seen in the methylamine, methanol, methane, and perfluoromethane cases is 0.33, 0.17, 0.002, and 0.002 e^-, respectively. Examining the UF_6-FG intermolecular distances, the propanamides are also able to approach the U atom slightly closer than the best N-only species (~2.4 vs. 2.5 Å). However, while the propanamides are able to distort the UF_6 molecule further than the best N-only species, the main interaction is between U and an O lone pair, which is less advantageous than interaction with N lone pairs; carbonyl oxygens were also seen in the O-only species to yield the lowest IBEs in the class. Taken together, these analyses suggest that strong IBE values in UF_6-FG complexes arise from a synergism of contributions between conformational deformation, electrostatic and vdW stabilizing interactions, and steric accessibility for U approach.

3.7 Correlation of UF_6-FG IBEs with the BF_3 Affinity Scale

Limited attention has been given recently to the Lewis acid behavior of UF_6 [35-39]. However, UF_6, and metallic fluorides in general, can act as Lewis acids, and could potentially be utilized in synthetic procedures as such. We hypothesized that our computed IBE values could be used as good surrogates for predicting Lewis acid/base behavior between UF_6 and the various functional groups. To this end, we have compared the IBE values with functional group molecules included within data establishing the BF_3 Affinity Scale used in labs to predict the Lewis behavior of ligands in synthetic pathways [40]. While Gal and co-workers have compiled hundreds of Lewis adduct complexation enthalpies over the past couple decades for a range of increasingly complex molecules, in our UF_6 screening study, we choose to use the smallest, most simple molecules that displayed various organic functional groups, which resulted in a rather few of our molecules being included in the database. Nevertheless, the initial comparison between the few overlapping chemical species shows a strong correspondence between our calculated UF_6 IBEs and the same functional group molecules' BF_3 affinity values, as seen in Figure 6.

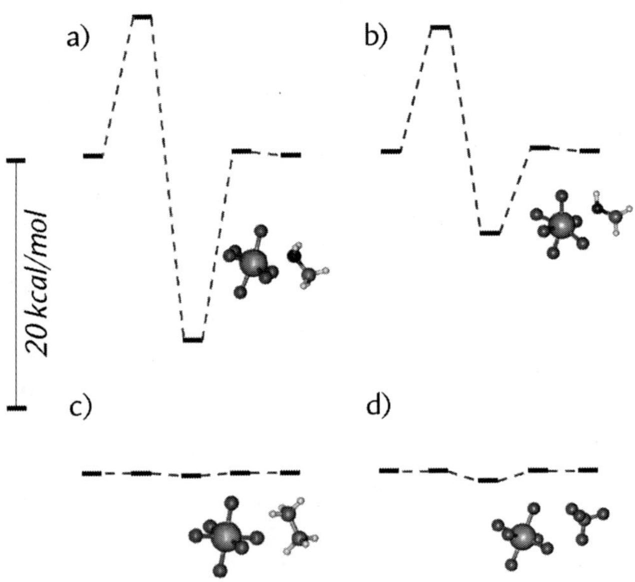

Figure 5. Energy diagrams for a range of UF_6-FG pairs. Each is set with the UF_6 on the left and the FG on the right, with their individual solo optimized geometries being set as zero. The IBE is in the middle, and the intermediate energies correspond to the lone molecules in the distorted geometry they assume in the UF_6-FG dimer. Often the UF_6 becomes significantly distorted to form the dimer bond. The panels correspond to UF_6 complexing with a) ethylamine, b) methanol, c) ethane, and d) perfluoromethane. Adapted with permission from [22].

The Figure displays good linear correlation between the two measures, but the examined molecules are quite simple. Further studies will elucidate whether deviations from the BF_3 scale will occur for UF_6 adducts for more complicated species due to the increased steric effects inherent in these UF_6 adducts. The BF_3 complexation enthalpy is sensitive to steric hindrance, as can be seen for bases with bulky substituents. It is possible the UF_6 may be even more sensitive, so it will be interesting to examine molecules in complexation with UF_6 that are more sterically complex. The BF_3 scale offers a large choice of small molecules, but in Gal's work it has not been possible to study experimentally compounds with O-H or N-H because of self-association or secondary reactions. These and other experimentally-difficult compounds can be approached with computational methods. Indeed, Gal and co-workers have made some efforts toward computing BF_3 enthalpies and have investigated the solvation effect on these values [41, 42]. The consideration of solvation and conformational effects is important for these types of calculations, as our lab has shown in other systems [43-46]. There are some classes of compounds that deviate in their Lewis behavior from the BF_3 scale – such as sulfur bases and unsaturated hydrocarbons; we did not investigate any of these types of molecules within this initial UF_6 IBE project, so it will also be interesting to see if these classes of compounds also deviate from the expected UF_6 Lewis adduct behavior or not.

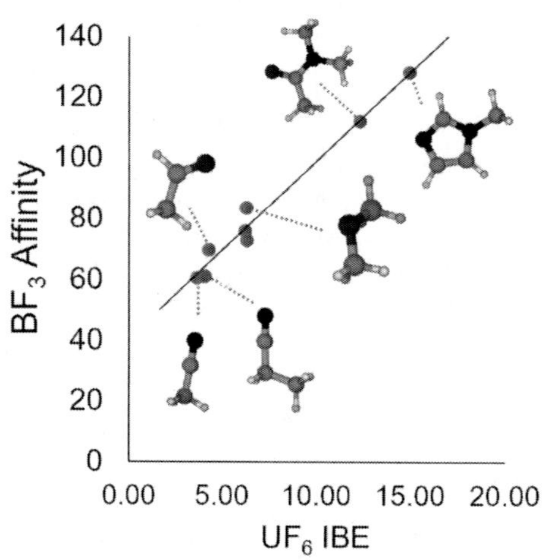

Figure 6. BF_3 Affinity values from Ref. [39] compared with UF_6 IBE values from Ref. [22] for common molecules examined in the two references. Molecular structures for most of the data points are inset in the figure.

Concluding Remarks

A leading goal of the studies described in this chapter was to screen functional group interactions with UF_6 to predict of how various surfaces, materials, or gases may interact with UF_6. Of the functional groups examined, the relatively structurally simple amine and imidazole FGs showed the highest IBE values, while hydrocarbons and perfluorocarbons showed the weakest UF_6 interactions. UF_6 seems to prefer interacting through lone pairs, with the singular N lone pairs performing better than O lone pairs. It appears that the geometric inconvenience of the second lone pair (a spectator to the main binding event) likely sterically clashes with the F atoms on UF_6 in a way that is not present with N species. UF_6 is only able to form strong IBE values when interacting principally with an O species if and when it experiences large geometric distortions to alleviate the steric clash. The nature of the UF_6-FG binding seems to depend on the oxidation state of interacting oxygen species, presumably by shifting the potential energetics of the would-be lone pair binding partner. Taken together, this work indicates that for strong IBE values in UF_6-FG complexes, there is a need to balance and optimize contributions between conformational deformation, electrostatic and vdW stabilizing interactions, and steric accessibility to approach of the uranium atom.

Finally, a subset of our UF_6-FG pairs were correlated with BF_3 affinity values to explore the potential for Lewis acidity. Overall, very good agreement between the "UF_6-IBE scale" and the BF_3 affinity scale are seen. Correlating the large body of data of BF_3 affinity values to UF_6-FG interactions may further help to identify where UF_6 may be useful for particular processes. We hope that information about the interactions of UF_6 with various functional groups may be of wide interest, especially insofar as it may inspire researchers to think about creative ways to make use of the large stores of depleted UF_6 "waste" from enrichment processes which we see as potentially very valuable untapped resource.

Acknowledgments

The authors are thankful to Jean-François Gal for discussions in support of this work. This work was funded by the United States National Nuclear Security Administration Office of Defense Nuclear Nonproliferation Research and Development. Preparation of the manuscript was supported by a Summer

Faculty Stimulus Grant from the College of Arts and Sciences at The University of Memphis. The High Performance Computing Center and the Computational Research on Materials Institute at The University of Memphis (CROMIUM) also provided generous resources for this research. Any opinions, findings, and conclusions or recommendations expressed in this book chapter are those of the author(s) and do not necessarily reflect the views of the funding/supporting agencies.

A large portion of this chapter is adapted from Charbonnet. K. A., Alexander, W. A. Computational Studies of Uranium Hexafluoride Interacting with Functionalized Organics. I. Screening Intermolecular Potentials between UF_6 and Small Organic Functional Groups. J. Phys. Chem. A 2021, 125, 10130-10137. With permission from American Chemical Society.

Figures in this chapter have undergone graphical alterations/edits to conform to publication standards and are intended for use in this chapter as *illustrations*. The original referenced figures should be used for rigorous scientific examination/comparison.

This work would not have been possible without the intellectual contributions of Dr. Megan E. Bennett.

References

[1] DOE, 2004, Final Environmental Impact Statement for Construction and Operation of a Depleted Uranium Hexafluoride Conversion Facility at the Portsmouth, Ohio, Site, DOE/EIS-0360. *US Department of Energy*, Office of Environmental Management.

[2] *Ports Virtual Museum.* "1991 cylinder storage area" digitized photograph. https://www.portsvirtualmuseum.org/images/photos/66-large.jpg.

[3] Schneider, B., Boring, A. M., Cohen, J. S. Interaction potentials for UF_6 with itself and with rare-gas atoms. *Chem. Phys. Lett.*, 1974, *27*, 577. https://doi.org/10.1016/0009-2614(74)80308-8.

[4] Beu, T. A., J. Onoe, J., Takeuchi, K. Homogenous and mixed UF_6 clusters with Ar: Calculations of structures and vibrational spectra. *J. Chem. Phys.*, 1998. *109*, 8295. https://doi.org/10.1063/1.477492.

[5] Beu, T. A., Onoe, J., Takeuchi, K. Calculations of structure and IR-spectrum for small UF_6 clusters. *J Chem. Phys.*, 1997, *106*, 5910. https://doi.org/10.1063/1.473256.

[6] Zarkova, L., Pirgov, P. Transport and equilibrium properties of UF_6 gas simultaneously fitted by an effective isotropic potential with temperature-dependent parameters. *J. Phys. B: At. Mol. Opt. Phys.*, 1995, *28*, 4269. https://doi.org/10.1088/0953-4075/28/19/012.

[7] Gagliardi, L., Willetts, A., Skylaris, C. K., Handy, N. C., Spencer, S., Ioannou, A. G., Simper, A. M. A relativistic density functional study on the uranium hexafluoride and plutonium hexafluoride monomer and dimer species. *J. Am. Chem. Soc.*, 1998, *120*, 11727. https://doi.org/10.1021/ja9811492.

[8] Hu, S. W., Liu, X. Q., Wang, X. Y., Chu, T. W. Effect of H_2O on the hydrolysis of UF_6 in the gas phase, *Journal of Molecular Structure*, 2014, *1062*, 29-34 (2014). https://doi.org/10.1016/j.molstruc.2014.01.015.

[9] Hu, S. W., Wang, X. Y., Chu, T. W., Liu, X. Q. Theoretical Mechanism Study of UF_6 Hydrolysis in the Gas Phase (II), *J. Phys. Chem. A*, 2009, *113*, 9243. https://doi.org/10.1021/jp904655w.

[10] Hu, S. W., Wang, X. Y., Chu, T. W., Liu, X. Q. Theoretical Mechanism Study of UF_6 Hydrolysis in the Gas Phase, *J. Phys Chem A*, 2008, *112*, 8877. https://doi.org/10.1021/jp804797a.

[11] Lind, M. C., Garrison, S. L., Becnel, J. M. Trimolecular reactions of uranium hexafluoride with water, *J. Phys. Chem. A*, 2010, *114*, 4641. https://doi.org/10.1021/jp909368g.

[12] Lu, J. W., Day, B. S., Fiegland, L. R., Bennett, M. E., Alexander, W. A., Lu, J. W., Troya, D., Morris, J. R. Collisions of Polar and Nonpolar Gases with Hydrogen Bonding and Hydrocarbon Self-Assembled Monolayers. *Journal of Physical Chemistry C*, 2008, *112*, 17272-17280. https://doi.org/10.1021/jp801476f.

[13] Alexander, W. A., Zhang, J., Murray, V. J., Nathanson, G. M., Minton, T. K. Kinematics and Dynamics of Atomic-Beam Scattering on Liquid and Self-Assembled Monolayer Surfaces, *Faraday Discussions*, 2012, *157*, 355-374. https://doi.org/10.1039/C2FD20034A.

[14] Day, B. S., Morris, J. R., Troya, D. Classical Trajectory Study of Collisions of Ar with Alkanethiolate Self-Assembled Monolayers: Potential-energy Surface Effects on Dynamics. *J. Chem. Phys.*, 2005, *122*, 214712. https://doi.org/10.1063/1.1924543.

[15] Alexander, W. A., Troya, D. Theoretical Study of the Dynamics of Collisions Between HCl and ω-Hydroxylated Alkanethiol Self-Assembled Monolayers. *J. Phys. Chem. C*, 2011, *115*, 2273- 2283. https://doi.org/10.1021/jp1106422.

[16] Alexander, W. A., Troya, D. Theoretical study of the Ar–, Kr–, and Xe–CH_4, –CF_4 intermolecular potential-energy surfaces. *J Phys Chem A*, 2006, *110*, 10834–10843. https://doi.org/10.1021/jp063398f.

[17] Lu, J. W., Day, B. S., Fiegland, L. R., Davis, E. D., Alexander, W. A., Troya, D., Morris, J. R. Interfacial energy exchange and reaction dynamics in collisions of gases on model organic surfaces. *Prog Surf Sci.*, 2012, *87*, 221–252. https://doi.org/10.1016/j.progsurf.2012.07.002.

[18] Alexander, W. A., Morris, J. R., Troya, D. Experimental and theoretical study of CO collisions with CH_3-and CF_3-terminated self-assembled monolayers. *J Chem Phys*, 2009, *130*, 084702. https://doi.org/10.1063/1.3072012.

[19] Lu, J. W., Alexander, W. A., Morris, J. R. Gas–surface energy exchange and thermal accommodation of CO_2 and Ar in collisions with methyl, hydroxyl, and perfluorinated self-assembled monolayers. *Phys Chem Chem Phys.*, 2010, *12*, 12533–12543. https://doi.org/10.1039/B921893A.

[20] Alexander, W. A., Day, B. S., Moore, H. J., Lee, T. R., Morris, J. R., Troya, D. Experimental and theoretical studies of the effect of mass on the dynamics of gas/organic-surface energy transfer. *J Chem Phys.*, 2008, *128*, 014713. https://doi.org/10.1063/1.2815327.

[21] Alexander, W. A. On the accuracy of analytical potentials: comment on 'Accurate ab initio calculation of the Ar–CF_4 intermolecular potential energy surface,' *Molecular Simulation*, 2015, *41*, 610-612. https://doi.org/10.1080/08927022.2014.907494.

[22] Charbonnet. K. A., Alexander, W. A. Computational Studies of Uranium Hexafluoride Interacting with Functionalized Organics. I. Screening Intermolecular Potentials between UF_6 and Small Organic Functional Groups. *J. Phys. Chem. A* 2021, *125*, 10130-10137. https://doi.org/10.1021/acs.jpca.1c05853.

[23] Kendall, R. A., Dunning Jr., T. H., Harrison, R. J. Electron Affinities of the First-row Atoms Revisited. Systematic Basis Sets and Wave Functions. *J. Chem. Phys.* 1992, *96*, 6796. https://doi.org/10.1063/1.462569.

[24] Peterson, K. A., Dunning, T. H. Accurate Correlation Consistent Basis Sets for Molecular Core–Valence Correlation Effects: The Second Row Atoms Al–Ar, and the First Row Atoms B–Ne Revisited. *J. Chem. Phys.* 2002, *117*, 10548–10560. https://doi.org/10.1063/1.1520138.

[25] Peterson, K. A. Correlation Consistent Basis Sets for Actinides. I. the Th and U Atoms. *J. Chem. Phys.* 2015, *142*, 074105. https://doi.org/10.1063/1.4907596.

[26] Dolg, M., Cao, X. Accurate Relativistic Small-Core Pseudopotentials for Actinides. Energy Adjustment for Uranium and First Applications to Uranium Hydride. *J. Phys. Chem. A* 2009, *113*, 12573-12581. https://doi.org/10.1021/jp9044594.

[27] Peterson, K. A., Figgen, D., Goll, E., Stoll, H., Dolg, M. Systematically convergent basis sets with relativistic pseudopotentials. II. Small-core pseudopotentials and correlation consistent basis sets for the post-d group 16–18 elements. *J. Chem. Phys.* 2003, *119*, 11113. https://doi.org/10.1063/1.1622924.

[28] Boys, S. F., Bernardi, F. The calculation of small molecular interactions by the differences of separate total energies. Some procedures with reduced errors. *Mol. Phys.*, 2002, *100*, 65-73. https://doi.org/10.1080/00268970110088901.

[29] Stephens, P. J., Devlin, F. J., Chabalowski, C. F., Frisch, M. J. Ab Initio Calculation of Vibrational Absorption and Circular Dichroism Spectra Using Density Functional Force Fields. *J. Phys. Chem.* 1994, *98*, 11623-11627. https://doi.org/10.1021/j100096a001.

[30] Grimme, S., Ehrlich, S., Goerigk, L. Effect of the damping function in dispersion corrected density functional theory. *J. Comp. Chem.* 2011, *32*, 1456-65. https://doi:10.1002/jcc.21759.

[31] Frisch, M. J., Trucks, G. W., Schlegel, H. B., Scuseria, G. E., Robb, M. A., Cheeseman, J. R., Scalmani, G., Barone, V., Petersson, G. A., Nakatsuji, et al., Gaussian 16, Revision B.01, Gaussian, Inc., Wallingford CT, 2016.

[32] Kaltsoyannis, N., Hay, P. J., Li, J., Blaudeau, J. P., Bursten, B. E. "Theoretical studies of the electronic structure of compounds of the actinide elements." in: *The Chemistry of the Actinide and Transactinide Elements*. Morss, L. R., Edelstein, N.

M., Fuger, J. (eds) Springer, Dordrecht. (2008). https://doi.org/10.1007/1-4020-3598-5_17.

[33] Graubner, T., Karttunen, A. J., Kraus, F., A Computational Study on Closed-Shell Molecular Hexafluorides MF_6 (M=S, Se, Te, Po, Xe, Rn, Cr, Mo, W, U) – Molecular Structure, Anharmonic Frequency Calculations, and Prediction of the NdF_6 Molecule. *ChemPhysChem*. 2023, *29*, e202200903. https://doi.org/10.1002/cphc.202200903.

[34] Aldridge, J. P., Brock, E. G., Filip, H., Flicker, H., Fox, K., Galbraith, H. W., Holland, R. F., Kim, K. C., Krohn, B. J., Magnuson, et al., Measurement and Analysis of the Infrared-active Stretching Fundamental (ν 3) of UF_6. *J. Chem. Phys*. 1985, *83*, 34–48. https://doi.org/10.1063/1.449777.

[35] Hunt, R. D., Andrews, L., Toth, L. M. Infrared Spectra of UF_6, WF_6, MoF_6, and SF_6 Complexes with Hydrogen Fluoride in Solid Argon. *J. Phys. Chem.* 1991, *95*, 1183-1188. https://doi.org/10.1021/j100156a028.

[36] Baluka, M., Edelstein, N., O'Donnell, T. A. Electronic Solution Spectra for Uranium and Neptunium in Oxidation States III to VI in Anhydrous Hydrogen Fluoride. *Inorg. Chem.* 1981, *20*, 3279-3283. https://doi.org/10.1021/ic50224a027.

[37] Hunt, R. D., Andrews, L., Toth, L. M. Infrared Spectra of HCl Complexes with UF_6, WF_6, and MoF_6. *Inorg. Chem.* 1991, *30*, 3829-3832. https://doi.org/10.1021/ic00020a011.

[38] Hunt, R. D., Sherrow, S. A. FTIR Spectra of the Hydrolysis of Uranium Hexafluoride. *J. Phys. Chem.* 1992, *96*, 1095-1099. https://doi.org/10.1021/j100182a015.

[39] Olivier Roy, O., Marquet, B., Alric, J. P., Jourdan, A., Morel, B., Langlois, B. R., Billard, T. (Re)Investigation of the reactivity of uranium hexafluoride toward several organic functions at room temperature, *Journal of Fluorine Chemistry*, 2014, *167*, 74-78. https://doi.org/10.1016/j.jfluchem.2014.06.007.

[40] Laurence, C., Gal, J. F., *Lewis basicity and affinity scales: data and measurement*. Wiley, 2009.

[41] Gal, J. F., Maria, P. C., Yáñez, M., Mó, O. On the Lewis Basicity of Phosphoramides: A Critical Examination of Their Donor Number through Comparison of Enthalpies of Adduct Formation with $SbCl_5$ and BF_3. *ChemPhysChem* 2019, *20*, 2566. https://doi.org/10.1002/cphc.201900691.

[42] Gal, J. F., Maria, P. C., Yáñez, M., Mó, O. Enthalpies of Adduct Formation between Boron Trifluoride and Selected Organic Bases in Solution: Toward an Accurate Theoretical Entry to Lewis Basicity. *Molecules* 2021, *26*, 6659. https://doi.org/10.3390/molecules26216659.

[43] Alexander, W. A. Toward Computational Accuracy in Realistic Systems to Aid Understanding of Field-Level Water Quality Issues. *Physchem* 2021, *1*, 243-249. https://doi.org/10.3390/physchem1030018.

[44] Alexander, W. A. Methyl-Cyclohexane Methanol (MCHM) Isomer-Dependent Binding on Amorphous Carbon Surfaces. *Molecules* 2021, *26*, 3411. https://doi.org/10.3390/molecules26113411.

[45] DeYonker, N. J., Charbonnet, K. A., Alexander, W. A. Dipole moments of trans- and cis-(4-methylcyclohexyl)methanol (4-MCHM): obtaining the right conformer

for the right reason. *Phys. Chem. Chem. Phys.*, 2016, *18*, 17856-17867. https://doi.org/10.1039/C5CP05952F.

[46] Dietrich, A. M., Thomas, A., Zhao, Y., Smiley, E., Shanaiah, N., Ahart, M., Charbonnet, K. A., DeYonker, N. J., Alexander, W. A., Gallagher, D. L. Partitioning, Aqueous Solubility, and Dipole Moment Data for *cis*- and *trans*-(4-Methylcyclohexyl)methanol, Principal Contaminants of the West Virginia Chemical Spill. *Environmental Science & Technology Letters* 2015, *2*, 123-127. https://doi.org/10.1021/acs.estlett.5b00061.

Section C

Environmental Implications

> My experiments proved that the radiation of uranium compounds can be measured with precision under determined conditions and that this radiation is an atomic property of the element of uranium.
>
> ***-Marie Curie***

Puzzle 3

Uranium on the Table

by rockratzero

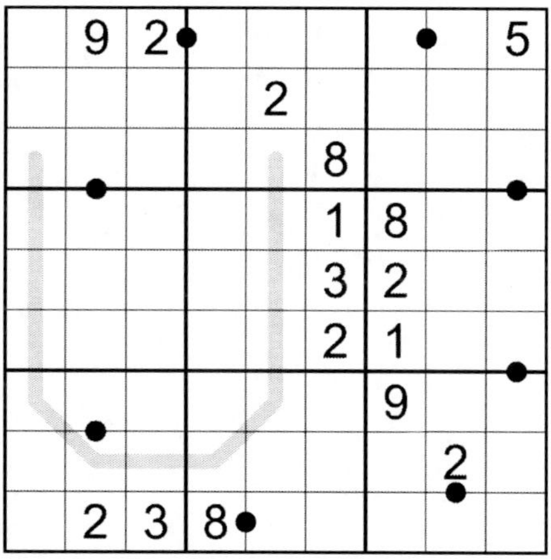

Rules

Place the digits 1-9 in each row, column, and 3x3 region without repeating.

Digits on the gray lines alternate between odd and even numbers.

Digits joined by black dots must be in a 2:1 ratio (one is double the other).

Link to SudokuPad online: https://tinyurl.com/rrz-uranium3.

Chapter 5

A Global Assessment of Uranium Concentrations in Water for Sustainability

Manoj Kumar Jindal[1]
Kiran Patni[2,*]
Ashutosh Pratap Pande[3]
Garima Punetha[4]
Chitra Pande[4]
Santosh Kumar Sar[5]
William A. Alexander[6]
and Priyanka Tiwari[7]
[1]Divecha Centre for Climate Change, Indian Institute of Science, Bangalore, India
[2]Graphic Era Hill University, Bhimtal Campus, Uttarakhand, India
[3]LSMGPG College, Pithoragarh, Uttarakhand, India
[4]DSB Campus, Kumaun University, Nainital, India
[5]Bhilai Institute of Technology, Durg, India
[6]Department of Chemistry, The University of Memphis, Memphis, TN, USA
[7]MBGPG College, Haldwani, Uttarakhand, India

Abstract

In recent years, several studies have been conducted to determine the potability of water in terms of uranium contamination. Uranium is a naturally occurring radioactive element, and its presence in water beyond the acceptable level can be considered extremely hazardous to human

[*] Corresponding Author's Email: 93kiranpatni@gmail.com

In: The Chemistry of Uranium
Editor: William A. Alexander
ISBN: 979-8-89530-802-8

health. Long-term intake of uranium-contaminated water has several health hazards, including kidney damage, genotoxicity, breast cancer, and cardiovascular disease. While water quality has historically been assessed using physicochemical parameters, more recently investigation of uranium and radionuclide content has gained attention due to the recognition of the potential for severe effects on human health. The World Health Organization (WHO) first advocated a limit of 15 μg/L, but this was later increased to 30 μg/L, while the Atomic Energy Regulatory Board (AERB) has set a tolerable limit of uranium in drinking water at 60 μg/L. Most natural water sources contain measurable levels of uranium. Uranium can be detected via laser fluorimetry, ICP-MS, atomic absorption, UV fluorimetry, gamma spectrometry, fission tracking, alpha spectrometry, liquid chromatography, liquid scintillation counting, and/or neutron activation methods. This chapter describes a global overview of the present scenario of uranium levels in water samples, risks associated with the higher detected concentration levels, and some suggestions for economically based remediation techniques. Water is ingested by all living things, thus if it is contaminated, it may have an impact on all living things, either directly or indirectly, affecting sustainability.

Keywords: Uranium, India, drinking water, human health, sustainability

1. Introduction

1.1. Water

Water is arguably the most important of our natural resources, having always been the key for all of Earth's living beings, as it is essential for their survival [1]. Although water an abundant surface material (covering three-quarters of the Earth's surface), less than 1% of this available water is fresh and usable [2]. More than half of all organisms are aquatic and live *in* this water, while most of the rest of the organisms are composed of large fractions of this water (for instance, ~70% of the human body by mass is water). Water is used for industrial purposes, irrigation purposes, and drinking purposes worldwide. It shapes the land and plays a significant role in regulating climate [3]. In the present scenario, unplanned urbanization and industrialization has played a key role in deteriorating the water quality [4]. Even with plentiful water *sources*, water quality, water scarcity, and water system maintenance are among the prime concerns facing humanity and the world in the future [5].

Water is considered polluted when it is not fit for human consumption and cannot support its natural biotic inhabitants. Natural water is getting depleted daily due to various impurities; the nature and amount of these impurites depends on the water source [6]. In addition to the many natural factors such as weathering of rocks, leaching of soils, and dissolution of aerosol particles from the atmosphere, additional strains on the watershed come from anthropogenic activities including mining and disposal of treated and/or untreated toxic waste; all these factors contribute to depleting water quality and resulting in the emergence of serious environmental problems [7]. Worldwide studies indicate that water pollution is the major cause of disease and death. According to the World Health Organization, the cause of about 80% of human disease can be attributed to issues with water quality [8]. Polluted water is linked to various negative health outcomes like cancer, reproductive complications, cardiovascular disease, dental problems, and neurological disease [9]. Water quality is directly related to human health and is deteriorating mainly due to human activities [4]. The availability of freshwater is decreasing with an increase in pollution. Pollution affects not only surface waters, such as lakes and rivers, but also groundwater; as a result, groundwater and impacted soils may become unusable for agriculture and irrigation purposes [10].

Groundwater is the largest source of freshwater after polar ice caps on the Earth. Groundwater plays a crucial role in the hydrologic cycle, plant growth, soil formation, and providing water for human activities. Since it is quite difficult to track groundwater, various hydrologic modelling, mapping, and monitoring are done to understand its distribution, storage, and transport through the subsurface. These monitoring activities are also essential to analyze and understand how groundwater can be used for a long time and can be protected from contaminants [11].

1.2 Natural Radioactivity in Water

The environmental toxicity of uranium is receiving increased attention due to the global push for low-carbon energy. This drive may increase reliance on nuclear power, leading to higher concentrations of depleted uranium nuclear wastes, alongside other natural and anthropogenic sources [12, 13]. Uranium occurs naturally in certain kinds of rocks, soils, and minerals. As water flows through these formations, uranium-containing compounds can dissolve and disperse. Anthropogenic activities, especially mining, increase the likelihood of uranium deposits being exposed to the water table, resulting in higher concentrations of uranium compounds in the water than would occur naturally.

Uranium compounds exhibit both chemical and radiological toxicity, with chemical hazards generally being more significant. This chemical toxicity is largely due to the non-biodegradable nature of these uranium compounds. Uranium affects the kidney, liver, and bones due to bioaccumulation and can cause severe health issues even with small exposures over long periods. Studies also indicate that uranium affects the respiratory and reproductive systems [14]. Therefore, heath organizations monitor concentrations uranium in drinking water. The World Health Organization recommends a limit of 30 $\mu g\ L^{-1}$ [15], while the Atomic Energy Regulatory Board of India sets a maximum acceptable level of 60 $\mu g\ L^{-1}$ [16]. The reference level is derived from epidemiological studies, based on the assumption of a 60 kg adult consuming 2 liters of drinking water per day and 80% allocation of the Tolerable Daily Intake (TDI) to drinking water.

Uranium is found in detectable concentrations in most natural water sources, existing in both dissolved and particulate forms [17]. While uranium can exist in both tetravalent and hexavalent forms, the hexavalent form is more water-soluble, while the tetravalent form is relatively insoluble [18]. Important minerals that are natural sources of uranium include pitchblende, silicates, vanadates, monazites, lignite sands, and uranium phosphates [17]. While uranium intake can occur through various routes, the main route of ingestion of uranium is through drinking water, followed to a lesser extent by foodstuffs. Uranyl ion forms bicarbonate, citrate, and $UO_3(CO_2)_3$ complexes in blood plasma. UO_2^{++} ions bind with red blood cells. During the blood purification in the kidney, uranium is filtered from the blood and recombined with cell surface ligands [19]. Uranium contamination in drinking water is increasing, not only in areas with uranium mines but also in regions where uranium occurs naturally [20, 21]. Continuous monitoring of these vulnerable water systems is essential to mitigate such conditions.

2. Review of Uranium Literature

2.1. Uranium in the Environment

Uranium is widely distributed in Earth's crust in the form of three isotopes: ^{238}U, ^{235}U, and ^{234}U. ^{238}U and ^{235}U are the parent nuclides of two decay series and ^{234}U results from the decay of ^{238}U [17]. Human beings are naturally exposed to radioactivity by ingesting and inhaling radionuclides present in food, water, and air. The World Health Organization previously set a limit of 15 μg/L in drinking water [22] but the limit has since been changed to 30 μg/L. This reference limit is based on epidemiological studies, as long-term exposure to uranium in water poses a potential health risk [15].

Studies have indicated that uranium contaminantion levels are lower in surface water than in groundwater [23, 24]. This difference arises because groundwater has direct contact with uranium-containing rocks and minerals and interacts with different geological formations compared to surface water. Specific geological formations that contain higher uranium concentrations may contribute to uranium contamination in surface water samples. The International Atomic Energy Agency (IAEA) classifies geological formations into fifteen types based on uranium deposits. This classification considers the fifteen types and their subtypes, along with countries where they are found. As of March 2, 2022, the IAEA has a total of 175 members worldwide, with most countries participating in the organization. Classification of geological formations (type, subtype, and class) as defined in the IAEA report [25] is shown in Table 1, based on worldwide uranium deposit sites.

The geological composition is a major factor influencing uranium deposits. Uranium contaminations in water depends on the geology of the surface and subsurface. Uranium does not readily dissolve into water upon immediate contact with rocks containing uranium. Instead, specific conditions and environmental factors, combined with prolonged contact times, are necessary to introduce uranium into the water table. Uranium is primarily found in the hexavalent state in water. Other uranium compounds are either sparingly soluble or insoluble.

2.2. Current Status of Uranium in Drinking Water

Numerous studies have investigated uranium levels in groundwater worldwide. According to the UNSCEAR report, the highest relative uranium concentrations are found in the human skeleton (57% of total uranium detected), followed by muscles, fat, bloodstream, lungs, liver and kidneys (20%, 16%, 4%, 2%, and 1%, respectively). The report also indicates that uranium levels in the human brain range from 0.4% and 0.99%. Concentration levels of uranium intake via drinking water, eating food, and inhalation vary between countries. Based on the UNSCEAR report, China has the highest uranium per-day dietary intake, according to data from twelve countries [23]. Many researchers have also investigated uranium levels at various locaions, and a few examples are included here.

Based on a study conducted to determine the uranium concentration in groundwater samples from Northern Greece, the concentration varied significantly, ranging from 0.15 μg/L to 7.66 μg/L. Springs exhibited higher uranium content than deep wells, with the exception of Apollonia spring (0.15 μg/L) [26].

Water samples of 129 wells were collected from seven of the nine subdivisions of Ulaanbaatar, Mongolia, and analyzed by inductively coupled plasma mass spectrometry (ICP-MS) using clean lab methods to quantify several trace elements, including As, Mn, Co, Zn, Se, Pb, and U. The levels of most of these trace elements was extremely low; however, uranium levels were surprisingly high, ranging from <0.01 to 57 μg/L, with a mean value of 4.6 μg/L. It was suggested that this elevated uranium content in the samples could be attributed to the local rocks and soil. This higher uranium concentration has been linked to nephrotoxicity, high blood pressure, bone dysfunction, and reproductive impairment in humans [27].

Another study examined 97 domestic water wells by the Apsaalooke (Crow) Tribal Research Group from Montana, USA. Two-thirds of the samples had detectable uranium concentrations, and 6.3% of samples had more than 30 μg/L [28].

Kopylova et al., conducted an experiment to assess the behavior of uranium and thorium in the "Choygan mineral water" natural spa area of Siberia. The region has 33 natural spring discharges that were sampled. The water type was mainly HCO_3-Na-Ca type. Total Dissolved Solids ranged from 300 to 2600 mg/L, the temperature ranged from 7 to 39°C, the pH ranged from 5.9 to 8.3, and the oxidation-reduction potential ranged from -170 to 236 mV. Uranium concentration varied from 0.7 to 14.0 μg/L. The study indicated that acid-alkaline and oxidizing- reducing conditions were the determining factors for uranium behavior and speciation in the studied water samples [29].

Abojassim and Mohammed conducted a study to compare the uranium concentrations in 40 tap water samples from the Al-Manathera and Al-Herra regions of Al-Nazaf, Iraq, using solid-state nuclear track detection (CR-39 detector) [17]. The average uranium concentrations at Al-Manathera were larger compared to Al-Herra regions but lower than the recommended values of 1.9μg/L set by the ICRP (International Commission on Radiological Protection) [30]. It was concluded that the water in that area is safe for use without any health risk [17].

Table 1. Geological classification of uranium deposit sites [25]

Type	Subtype and class	Site	Grade (% U)
1. Intrusive	1.1. Anatectic: pegmatite–alaskite 1.2. Plutonic: Class 1.2.1. Quartz monzonite Class 1.2.2. Peralkaline complex Class 1.2.3. Carbonatite	Rössing (Namibia), Bancroft district (Canada). Bingham Canyon (USA), Chuquicamata (Chile). Kvanefjeld (Greenland), Poços de Caldas (Brazil). Phalabora (South Africa), Catalão (Brazil).	Low to medium grade (0.01 to 0.10%)
2. Granite-related	2.1. Endogranitic 2.2. Perigranitic	La Crouzille district (France), Xiazhuang district (China). Příbram region (Czech Republic), Niederschlema (Germany).	Low to high grade (0.05 to 1%)
3. Polymetallic iron oxide breccia complex	-	Olympic Dam & Carrapateena (Australia).	Low grade (0.003 to 0.023%)
4. Volcanic-related	4.1. Stratabound 4.2. Structure-bound 4.3. Volcano-sedimentary	Dornod No. 7 ore zone (Mongolia), Maureen (Australia). Streltsov–Antei (Russian Federation), Kurišková (Slovakia). Anderson Mine (USA), Sierra Pintada District (Argentina).	Low to medium grade (0.03 to 0.40%)

Table 1. (Continued)

Type	Subtype and class	Site	Grade (% U)
5. Metasomatite	5.1. Na- metasomatite: Class 5.1.1. Granite derived Class 5.1.2. Metasediment–metavolcanic derived. 5.2. K-metasomatite 5.3. Skarn	Kirovograd district (Ukraine), Lagoa Real (Brazil). Krivoy Rog district (Ukraine). Elkon district (Russian Federation). Mary Kathleen (Australia), Tranomaro (Madagascar).	Low to medium grade (0.05 to 0.15%)
6. Metamorphite	6.1. Stratabound 6.2. Structure-bound: Class 6.2.1. Monometallic veins Class 6.2.2. Polymetallic veins Class 6.2.3. Marble-hosted phosphate.	Forstau (Austria), Nuottijarvi (Finland). Schwartzwalder (USA), Ace–Fay–Verna (Canada). Shinkolobwe (Democratic Republic of Congo). Itataia (Brazil), Zaozernoye (Kazakhstan).	Low to medium grade (0.05 to 0.50%)
7. Proterozoic unconformity	7.1. Unconformity-contact 7.2. Basement-hosted 7.3. Stratiform fracture-controlled	Cigar Lake, Key Lake & McArthur River (Canada). Jabiluka & Ranger (Australia), Eagle Point (Canada). Lambapur & Chitrial (India).	Low to high grade (0.05 to 20%)
8. Collapse breccia pipe	-	Arizona Strip (USA).	High grade (0.20 to 0.80%)

Type	Subtype and class	Site	Grade (% U)
9. Sandstone	9.1. Basal channel 9.2. Tabular: Class 9.2.1. Continental fluvial, U associated with intrinsic reductant Class 9.2.2. Continental fluvial, U associated with extrinsic bitumen Class 9.2.3. Continental fluvial vanadium–uranium. 9.3. Roll-front: Class 9.3.1. Continental basin, U associated with intrinsic reductant Class 9.3.2. Continental to marginal marine, U associated with intrinsic reductant Class 9.3.3. Marginal marine, U associated with extrinsic reductant. 9.4. Tectonic-lithologic 9.5. Mafic dykes/sills in Proterozoic sandstone	Dalmatovskoye (Russian Federation), Beverley (Australia). Arlit District (Niger), Ambrosia Lake District (USA). Arlit type (Niger). Grants type (USA). Salt Wash type (USA). Wyoming Province (USA), Chu-Sarysu Basin (Kazakhstan). Wyoming type (USA). ChuSarysu type (Kazakhstan). South Texas type (USA) Lodève Basin (France), Franceville Basin (Gabon). Westmoreland district (Australia), Matoush (Canada).	Low to high grade (0.01 to 3%)
10. Palaeo quartz-pebble conglomerate	10.1. U dominant 10.2. Au dominant	Elliot Lake district (Canada) Witwatersrand Basin (South Africa).	Low to medium grade (0.03 to 0.10%)

Table 1. (Continued)

Type	Subtype and class	Site	Grade (% U)
11. Surficial	11.1. Peat bog 11.2. Fluvial valley 11.3. Lacustrine–playa 11.4. Pedogenic and fracture fill	Kamushanovskoye (Kyrgyzstan), Flodelle Creek (USA). Yeelirrie (Australia), Langer Heinrich (Namibia). Lake Maitland & Lake Way (Australia). Beslet (Bulgaria).	Low to medium grade (0.01 to 0.10%)
12. Lignite–coal	12.1. Stratiform 12.2. Fracture-controlled	Koldzhat (Kazakhstan), Williston Basin (USA). Freital (Germany), Turakavak (Kyrgyzstan).	Low to medium (0.01 to 0.10%)
13. Carbonate	13.1. Stratabound 13.2. Cataclastic 13.3. Palaeokarst	Tumalappalle (India). Mailuu-Suu (Kyrgyzstan), Todilto district (USA). Sanbaqi (China), Tyuya-Muyun (Kyrgyzstan).	Low to high (0.03 to >1%)
14. Phosphate	14.1. Organic phosphorite 14.2. Minerochemical phosphorite 14.3. Continental phosphate	Mangyshlak Peninsula (Kazakhstan), Ergeninsky region (Russian Federation). Phosphoria Formation (USA). Bakouma district (Central African Republic)	Low grade (0.005 to 0.08%)
15. Black shale	15.1. Stratiform 15.2. Stockwork	Ranstad (Sweden), Chattanooga Shale Formation (USA). Ronneburg district (Germany), Dzhantuar (Uzbekistan).	Low to medium grade (0.005 to 0.20%)

Groundwater samples were collected from the German Federal State of Bavaria to analyze uranium concentration and incidence rates of various diseases related to increased uranium concentration (liver, musculoskeletal, tumors/growths, genito-urinary diseases, thyroid, and congenital malformations). A weak but significant positive correlation was found between uranium concentration and incidence rates of tumors/growths and liver disease. An increased rate of thyroid issues were diagnosed where mean uranium content in the drinking water was >2 μg/L [31].

Table 2. Uranium level ranges detected in water samples from few countries [23, 32]

Name of the place (city/country)	Number of analyzed samples	Range of the Uranium concentration (μg/L)
Finland	1058 g 336 s	0.001 to 3410 g 0.01 to 34 s
Sweden	481	<0.20 to 470 g
Poland	51	0.2 to 2.4 g
Spain	37	<0.41 to 56.1 g
Switzerland	5548	0.05 to 92.02 g
Greece	10	0.15 to 7.66 g
Canada	339	<0.02 to 2020 g
USA	55479 g 34561 s	0.03 to 7780 g 0.03 to 1737 s
Brazil	436	0.008 to 15.0 g
Morocco	15	0.37 to 25.1 g
Ethiopia	138	0.005 to 48 g
Egypt	12	1.19 to 519.4 g
Ghana	195	<0.001 to 1.99 g
Fujian Province, China	662	0.03 to 13.4 g
Iran	27 g 22 s	0.24 to 5.4 g 2.1 to 13.6 s
Bangladesh	67	0.2 to 10 g
Japan	23	0.0005 to 0.03 g
India	55554	≤0.2 to 4918 g ≤0.2 to 22 s
Argentina	92	<0.01 to 50 s

g = groundwater, s = surface water.

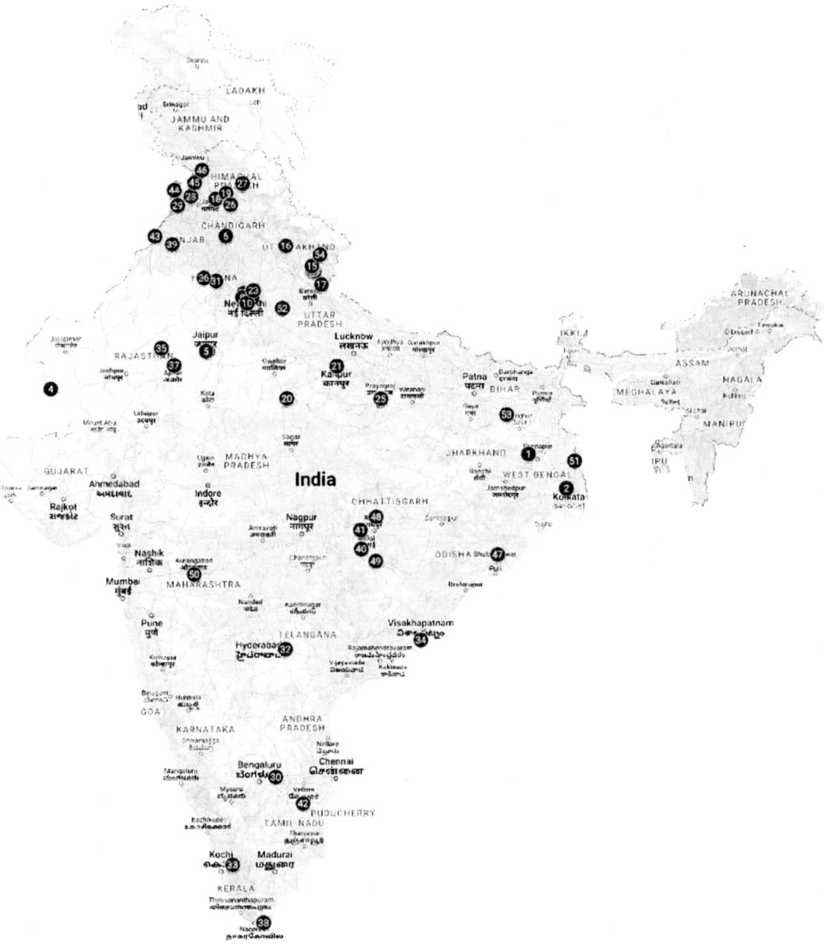

Figure 1. Approximate water sampling site locations at various locations/states in India. Numbered circles reference the entries in Table 3. Due to scale of the image, the circles overlap for locations 3 and 5; 7-10; and 11-15. Base map from Google Maps.

Aside from the data in Table 2, researchers and scientists have observed unexpected fluctuations in uranium content in water, which may vary depending on geography and other factors. For better understanding, India is taken as an example for information and explanations about uranium fluctuation among locations, based on published data by various researchers. India is one of the world's most populated countries, with a widely variable natural and atmospheric climate between states or cities. The available data regarding detected uranium water concentrations from different state-wide locations are compiled in Table 3. The geographic coverage of these locations is diverse and covers the majority of India's landmass, as can be appreciated in the map of (approximate due to scale) sampling locations seen in Figure 1.

The concentration of uranium was measured in groundwater samples collected from hand pumps in the Bhiwani, Hisar, Fatehabad, and Sirsa districts of Western Haryana, India. Uranium concentrations in the samples varied from 6.37 μg/L to 43.31 μg/L, with a mean of 19.14 μg/L. The investigation found that all the samples had uranium contents that were higher than the acceptable threshold of 1.9 μg/L [30]. However, most of the results were found to be equivalent to the WHO's safe level of 15 μg/L. Only 15% of the samples had concentrations above the recommended safe level of 30 μg/L. According to reports, the elevated uranium levels might be attributable to the area's radioactive-rich granites. Because 85 percent of the samples were found to be under the acceptable level, the people may not face any serious health risks [41].

A study was conducted to determine the amount of natural uranium in groundwater near the Jaduguda uranium mining complex in India. At the sample site (1.5 to 5 km distant zone), the maximum uranium content was observed to be 28 μg/L in the mining industry, which was lower than the WHO-recommended limit. The water near the mining sector was determined to be safe in three zones (1.5 km, 1.5 km-5 km, and >5 km), indicating a natural distribution of uranium in water [19].

Uranium was detected in 346 drinking water samples collected from the south coast districts of Kerala, which have extremely high natural background radiation due to monazite beach sand. The uranium concentration in the water ranged from 0.31-4.92 μg/L, below the recommended limits [43].

In Visakhapatnam, India, research was conducted to evaluate the uranium content in water samples for health risk assessment. The uranium concentration in the samples ranged from 0.6 to 12.3 μg/L, with none of the samples exceeding the limit of 30 μg/L [44].

The uranium concentration in Northern Rajasthan was measured using an inductively coupled plasma mass spectrometer (ICP-MS). Samples were collected from hand pumps in four districts: Hanumangarh (10 samples), Sri Ganganagar (10 samples), Churu (6 samples), and Sikar (13 samples). According to the study, the uranium content in the collected samples ranged from 2.54 to 133 μg/L, with a mean value of 38.48 μg/L. The majority of the samples were unsafe for consumption because they exceeded the safe limit [45].

Table 3. Uranium concentrations in drinking water samples at various locations/states of India

No.	Location	State	Water source	Uranium concentration (μg/L)
1	Asansol	West Bengal	Tap water	1.3 ± 0.03 [33]
2	Howrah	West Bengal	Tap water	13.2 ± 0.08 [33]
3	Galtaji, Jaipur	Rajasthan	Tap water	0.9 ± 0.01[33]
4	Jaisingder	Rajasthan	Tap water	2.5 ± 0.04 [33]
5	Central Jaipur	Rajasthan	Tap water	3 ± 0.08 [33]
6	Kauli	Punjab	Tap water	1.4 ± 0.02 [33]
7	East Delhi	Delhi	Tap water	0.6 ± 0.02, 0.9 ± 0.03 [33]
8	IGPI, Delhi	Delhi	Tap water	1.1 ± 0.03 [33]
9	Palam airport, Delhi	Delhi	Tap water	1.3 ± 0.02 [33]
10	South Delhi	Delhi	Tap water	1.3 ± 0.02 [33]
11	Bhimtal	Uttarakhand	-	1.66 ± 0.02 [34]
12	Nainital	Uttarakhand	-	1.82 ± 0.02 [34]
13	Kainchi Dham	Uttarakhand	-	2.32 ± 0.02 [34]
14	Rati Ghat	Uttarakhand	-	2.75 ± 0.03 [34]
15	Khairna	Uttarakhand	-	4.38 ± 0.03 [34]
16	Panyali	Uttarakhand	-	3.03 ± 0.03 [34]
17	Nanak Matta	Uttarakhand	-	5.46 ± 0.04 [34]
18	Samur Kalan	Himanchal Pradesh	-	18.20 ± 0.07 [34]
19	Rameda	Himanchal Pradesh	-	35.83 ± 0.09 [34]
20	Jhansi	Uttar Pradesh	Ground water	0.9-6.4 [35]
21	Kanpur	Uttar Pradesh	Ground water	3.3-9.1 [35]
22	Allahabad	Uttar Pradesh	Ground water	0.9-2.3 [35]
23	Ghaziabad	Uttar Pradesh	Ground water	4.2-11.4 [35]
24	Baluaghat	Uttar Pradesh	Yamuna river	0.92 [35]
25	Gaughat	Uttar Pradesh	Yamuna river	1.60 [35]
26	Bilaspur	Himachal Pradesh	Spring water	0.1-4.6 [36]
27	Kulu	Himachal Pradesh	Ground water	0.3-2.5 [37]
28	Amritsar	Punjab	Ground water	3.2-45.6 [38]

No.	Location	State	Water source	Uranium concentration (μg/L)
29	Malwa	Punjab	Ground water	5.4-43.4 [39]
30	Kolar	Karnataka	Ground water	0.3-1442.9 [40]
31	Western Haryana	Haryana	Ground water	6.37-43.31 [41]
32	Hyderabad	Telangana	Ground water	0.6-82.0 [42]
33	Kerala	Kerala	-	0.31-4.92 [43]
34	Vishakhapatanam	Andhra Pradesh	-	0.6-12.3 [44]
35	Northern Rajasthan	Rajasthan	Ground water	2.54-133 [45]
36	Western Haryana	Haryana	Ground water	0.3-256.4 [46]
37	Jaipur, Azmer	Rajasthan	Ground water	0.4-123.9 [47]
38	Kudankulam	Tamil Nadu	-	0.2-6.6 [48]
39	SW- Punjab	Punjab	Ground water	0.5-579 [49]
40	Balod	Chhattisgarh	Ground water	0.56-78.93 [50]
41	Durg	Chhattisgarh	Ground water	0.64-45.7 [51]
42	Tiruvannamalai	Tamil Nadu	-	1.9-23.4 [52]
43	Fazika	Punjab	Groundwater	4.32-83.99 [53]
44	Amritsar	Punjab	Ground water	0.6-65.3 [54]
45	Gurdaspur	Punjab	Ground water	0.1-38.8 [54]
46	Pathankot	Punjab	Ground water	0.1-18.5 [54]
47	Cuttack	Odisha	Ground water	0.1-6.5 [55]
48	Bemetara	Chhattisgarh	Ground water	0.68-96.08 [56]
49	Kanker	Chhattisgarh	Ground water	0.5-87.9 [56]
50	Beed, Aurangabad	Maharashtra	Ground water	0.03-32.85 [57]
51	Nadia	West Bengal	Ground water	0.21-20.9 [58]
52	Gunnaur	Uttar Pradesh	Ground water	0.11-39.76 [59]
53	Jamui	Bihar	Ground water	0.5-20.07 [60]
54	Bageshwar	Uttarakhand	Spring water	0.1-28.4 [61]
55	Nalbari	Assam	Ground water	0.3-10.3 [62]
56	Haryana	Haryana	-	0.65- 32.71 [63]
57	Kerala	Kerala	-	<0.5-12.54 [64]
58	Pithoragarh	Uttarakhand	Spring water	0.1-9.9 [65-6]
59	Eastern Karnataka	Karnataka	Ground water	1-5995 [67]

In the Uttar Pradesh districts of Khlilabad, Gorakhpur, Maharajganj, and Kushinagar, the uranium content in drinking groundwater samples was determined using laser fluorimetry. The study also calculated a cumulative, age-dependent radiation exposure to humans. The uranium content ranged from 0.20 μg/L to 64.0 μg/L, with an average of 11.1 μg/L. The quantity of uranium in water samples increased with depth, with 90% of samples below the WHO standard of 30 μg/L and only just 2% above the limit set by the AERB in India (60 μg/L) [68].

The uranium content in the drinking water of the Jaipur and Ajmer districts of Rajasthan was tested for exposure assessment. Residents in the research region consumed uranium in amounts ranging from 0.4 μg/L to 123.9 μg/L [47].

Research was conducted in 2016 to evaluate the uranium content in groundwater near Kudankulam in Southern Tamil Nadu. pH, EC, TDS, salinity, chloride, and sulphate were all measured. The uranium concentration in the water ranged from 0.2 to 6.6 μg/L, with a mean value of 2 μg/L. Uranium showed a positive relationship with TDS, EC, salinity, chloride, and sulphate, with natural uranium distributed equally in all zones within the studied range, indicating that natural uranium is prevalent [48].

Research was undertaken in the Southwest Punjab districts of India to examine the presence of uranium in 498 drinking water samples. The uranium content in the water samples ranged from 0.5 to 579 μg/L. The findings indicated that 338 out of 498 samples had uranium concentrations exceeding the WHO recommended limit of 30 μg/L, and 216 samples above the AERB level of 60 μg/L [49].

The presence of uranium in groundwater in the Punjab districts of Faridkot and Muktsar was determined using stable isotopes in water. Twenty samples were obtained in all, 10 of which were found to be contaminated. According to the findings, uranium concentrations varied from 3 to 190 μg/L. High uranium levels were observed in 44.4% and 81.8% of samples from Muktsar and Faridkot, respectively, and the percentage decreased as the depth of the sample increased [69].

A preliminary investigation examined uranium levels in drinking water from the Tiruvannamalai area of Tamil Nadu, India. Uranium levels ranged from 1.9 to 23.4 ppm, with a mean value of 6.94 ppm, and were determined to be below the permitted limits set by AERB, USEPA, WHO, UNSCEAR, and ICRP. The study concluded that the presence of uranium in the water samples from this area did not pose an increased lifetime cancer risk or a chemical health risk [52].

An investigation of groundwater samples from twenty villages in the Fazika district of Punjab, India revealed that uranium content ranged from 4.32 to 83.99 μg/L, with a mean of 26.51 μg/L. The study found that 24% of water samples exceeded the WHO criteria for safe drinking water, while 9%

exceeded the AERB level. The study suggested that the reion's geography was a key factor contributing to the elevated uranium content in the water samples [53].

Das et al., [58], Kumar et al., [61], and Patni et al., [65-66] found uranium content below prescribed limits in water samples collected from the West Bengal region, the Bageshwar region of Uttarakhand, and the Pithoragarh region of Uttarakhand, respectively, indicating that the water was safe for drinking.

Srinivasan et al., reported the highest uranium levels in the eastern part of Karnataka India [67]. That study of groundwater from 73 villages in Karnataka, India, indicated that 78% and 68% of the village samples showed uranium concentrations exceeding the WHO and AERB permitted levels of 30 μg/L and 60 μg/L, respectively. The highest uranium concentration was reported in Chikkevaripalli village (5995 ± 7 μg/L) [67].

3. Implications for Sustainability

3.1. Associated Health Hazards

Uranium exists in various valence states (+2, +3, +4, +5 and +6), but it is most commonly found in the hexavalent state in water due to its high solubility. It is rarely found in the tetravalent state in water. The health effects of uranium exposure can be categorized into chemical toxicity and radiological toxicity.

The primary chemical toxicity of uranium exposure involves kidney damage. Uranium can enter the body through inhalation of dust, or ingestion of contaminated food or liquids. It then circulates in the bloodstream and affects kidney cells. High concentrations of uranium intake can lead to kidney failure and mortality. Even low concentrations can affect the human system, but the body's repair mechanisms can usually manage the effects of low intake. However, at higher concentrations, the body's repair mechanisms become overwhelmed, with the threshold depending on the individual [70].

Uranium and its decay products release radiation, potentially causing a range of health effects from headaches to mortality, depending on the exposure level. Even low levels of radiation exposure can increase the risk of cancer over time, while high doses are linked to cancers through radiation-induced cell mutation. The probability of cancer increases with the dose and uptake of uranium.

Studies across various countries reveal differing concentration ranges of uranium in water, leading to a lack of universal standard for tolerable uranium intake. While the World Health Organization suggests a limit of 30 μg/L worldwide, individual countries have established their own values, both higher and lower. For example, Australia sets a limit of17 μg/L, Brazil at 30 μg/L, Bulgaria at 60 μg/L, Canada at 20 μg/L, the Czech Republic at 964 μg/L, Finland at 100 μg/L, Germany at 10 μg/L, India at 60 μg/L, the Russian Federation at 1700 μg/L, Slovenia at 6.8 μg/L, Slovakia at 346 μg/L, Sweden at 15 μg/L, Switzerland at 30 μg/L, the United States at 30 μg/L, and Japan at 2 μg/L [18, 24]. This illustrates a wide variation, from the stringent guidance in Japan to the highest levels permitted in the Russian Federation. Each country has a responsibility to determine these limits based on available data, aiming to prevent observed adverse effects. However, the wide range suggests uncertainty for this safe limit and warrants further investigation to better understand whether these variations stem from scientific approaches or other factors.

3.2. Drinking Water Sustainability and Economic Remediation

Radioactive materials in drinking water or inhaled via vapors pose health hazards. The presence of uranium in water sources is confirmed by various scientific organizations. Uranium concentrations in these sources often exceed safe limits set by regulatory authorities. Therefore, continued research into economical remediation techniques is essential to minimize uranium levels in water and ensure safe drinking water for all. Recent reviews detail research into cost-effective methods for controlling uranium levels.

Currently, individuals and families have limited options for purifying uranium-contaminated water. The primary recourse is often finding alternative water sources, potentially forcing them to purchase drinking water from other systems or locales. While private treatment facilities are possible, they are generally to expensive. Therefore, the responsibility for providing radionuclide-free drinking water largely falls on government water treatment facilities to control contaminants, including uranium, before distribution.

Various techniques exist for treating uranium-contaminated water, including precipitation, ion exchange, solvent extraction, nanofiltration, reverse osmosis, sorption, and liquid membrane separation. Activated carbon sorption and reverse osmosis are commonly used in practice, with the choice depending on the economics of materials and operation costs [71-72]. A review by Katsoyiannis and Zouboulis [73] discusses methods for removing uranium from drinking water, including coagulation and lime softening. The effectiveness of reactive sorption, absorption, or adsorption techniques is highly dependent on the properties of the sorptive materials used; an extensive review of many of these materials can be found within the report of Kausar and Bhatti [74], and solid phase sorbents for sorption and ion exchange methods are reviewed by Rosenberg et al., [75]. Liquid membrane extraction techniques have also been explored [75].

Simply put, drinking water is supplied through two major sources: groundwater and surface water. Groundwater generally has a higher uranium content, making its treatment essential, especially where uranium levels exceed the threshold limits recommended by WHO or individual countries. Because effective removal processes can be costly, ongoing research continues the search for economical routes for uranium removal.

Concluding Summary

Uranium concentration in water is a significant area of research concerning radioactive elements. Assuring water quality in terms of uranium contamination is crucial for achieving the United Nations Sustainable Development Goals. Uranium levels in drinking water should be below the limits set by WHO or the controlling authorities in each country. Most countries appear to have established their own uranium standard limit, based on water climate conditions, the presence of radionuclides in their geology, and other factors. The WHO guideline limit is 30 μg/L, but national standards range from 2 to 1700 μg/L. Uranium concentrations vary widely around the world, depending on the local geology, with some locations reporting levels below detection limits and others exceeding 5000 ug/L. Research into uranium in drinking water sources is essential to minimize the health impact of radioactive elements. Existing remediation measures should be employed when uranium levels exceed allowed limits.

References

[1] Asadi, E., Isazadeh, M., Samadianfard, S., Ramli, M. F., Mosavi, A., Nabipour, N. Shamshirband, S., Hajnal, E., Chau, K. W. Groundwater quality assessment for

sustainable drinking and irrigation. *Sustainability*, 2020, *12*(177). https://doi.org/10.3390/su12010177.

[2] Khatri, N., Tyagi, S. Influences of natural and anthropogenic factors on surface and groundwater quality in rural and urban areas. *Frontiers in Life Science*, 2015, *8*(1), 23–39. https://doi.org/10.1080/21553769.2014.933716.

[3] Sayemuzzaman, M., Ye, M., Zhang, F., Zhu, M. Multivariate statistical and trend analyses of surface water quality in the central Indian river Lagoon area, Florida. *Environmental Earth Sciences*, 2018, *127*(4), 1–13. https://doi.org/10.1007/s12665-018-7266-0.

[4] Wu, J., Sun, Z. Evaluation of Shallow Groundwater Contamination and Associated Human Health Risk in an Alluvial Plain Impacted by Agricultural and Industrial Activities, Mid-west China. *Exposure and Health*, 2016, *8*(3), 311–329. https://doi.org/10.1007/s12403-015-0170-x.

[5] Tiwari, K., Goyal, R., & Sarkar, A. GIS-Based Spatial Distribution of Groundwater Quality and Regional Suitability Evaluation for Drinking Water. *Environmental Processes*, 2017, *4*(3), 645–662. https://doi.org/10.1007/s40710-017-0257-4.

[6] Joshi, D. M., Bhandari, N. S., Kumar, A., Agrawal, N. Statistical analysis of physicochemical parameters of water of River Ganga in Haridwar district. *Rasayan Journal of Chemistry*, 2009, *2*(3), 579–587.

[7] Patil, P. N., Sawant, D. V., Deshmukh, R. N. Physico-chemical parameters for testing of water – A review. *International Journal of Environmental Sciences*, 2012, *3*(3), 1194–1207.

[8] Alirezazadeh, N., Garshasbi, N. A survey of natural uranium concentrations in drinking water supplies in Iran. *Iranian Journal of Radiation Research*, 2003, *1*(3), 139-142.

[9] Tareen, A., Sultan, I., Parakulsuksatid, P. Detection of heavy metals (Pb, Sb, Al, As) through atomic absorption spectroscopy from drinking water of District Pishin, Balochistan, Pakistan. *International Journal of Current Microbiology and Applied Science*, 2014, *3*(1), 299–308.

[10] Lone, S. A., Bhat, S. U., Hamid, A., Bhat, F. A., Kumar, A. Quality assessment of springs for drinking water in the Himalaya of South Kashmir, India. *Environmental Science and Pollution Research*, 2020, *28*(2), 2279–2300. https://doi.org/10.1007/s11356-020-10513-9.

[11] Brands, E., Rajagopal, R., Eleswarapu, U., Li, P. Groundwater. In *International Encyclopedia of Geography: People, the Earth, Environment and Technology*, 2017, (eds. D. Richardson, N. Castree, M. F. Goodchild, A. Kobayashi, W. Liu, and R. A. Marston). Wiley. https://doi.org/10.1002/9781118786352.wbieg0677.

[12] Agency for Toxic Substances and Disease Registry (ATSDR). (2013). Toxicological Profile for Uranium. Atlanta, GA: U.S. Department of Health and Human Services, Public Health Service. https://www.atsdr.cdc.gov/toxfaqs/tfacts150.pdf.

[13] Melo, D., Burkart, W. Uranium: Environmental Pollution and Health Effects. In *Encyclopedia of Environmental Health, 2nd Ed*, 2011, (ed. J. Nriagu). Elsevier. 252-258. https://doi.org/10.1016/B978-0-444-63951-6.00658-6.

[14] Limson Zamora, M., Tracy, B. L., Zielinski, J. M., Meyerhof, D. P., Moss, M. A. (1998). Chronic ingestion of uranium in drinking water: A study of kidney bioeffects

in humans. *Toxicological Sciences*, 1998, *43*, 68–77. https://doi.org/10.1006/toxs.1998.2426.

[15] World Health Organization. (2011). Guidelines for drinking-water quality, 4th ed. World Health Organization. https://apps.who.int/iris/handle/10665/44584.

[16] Atomic Energy Regulatory Board. (2004). Atomic Energy (Radiation Protection) Rules. https://aerb.gov.in/images/PDF/RPR2004-with-corrigendum.pdf.

[17] Abojassim, A. A., & Mohammed, H. A.-U. Comparing of the uranium concentration in tap water samples at Al-Manathera and Al-Herra Regions of Al-Najaf, Iraq. *Karbala International Journal of Modern Science*, 2017, *3*, 111–118. https://doi.org/10.1016/j.kijoms.2017.07.001.

[18] Balaram, V., Rani, A., Rathore, D. P. S. Uranium in groundwater in parts of India and world: A comprehensive review of sources, impact to the environment and human health, analytical techniques, and mitigation technologies. *Geosystems and Geoenvironment*, 2022, 1(2). https://doi.org/10.1016/j.geogeo.2022.100043.

[19] Sethy, N. K., Tripathi, R. M., Jha, V. N., Sahoo, S. K., Shukla, A. K., Puranik, V. D. Assessment of Natural Uranium in the Ground Water around Jaduguda Uranium Mining Complex, India. *Journal of Environmental Protection*, 2011, *2*(7), 1002–1007. https://doi.org/10.4236/jep.2011.27115.

[20] Arzuaga, X., Rieth, S. H., Bathija, A., Cooper, G. S. Renal effects of exposure to natural and depleted uranium: A review of the epidemiologic and experimental data. *Journal of Toxicology and Environmental Health*, 2010 *13*, 527–545. https://doi.org/10.1080/10937404.2010.509015.

[21] Kurttio, P., Auvinen, A., Salonen, L., Saha, H., Pekkanen, J., Mäkeläinen, I., Väisänen, S. B., Penttilä, I. M., Komulainen, H. Renal effects of uranium in drinking water. *Environmental Health Perspectives*, 2002, *110*(4), 337–342. https://doi.org/10.1289/ehp.02110337.

[22] World Health Organization. (2004). Uranium in drinking-water: background document for development of WHO guidelines for drinking-water quality. World Health Organization. https://apps.who.int/iris/handle/10665/75365.

[23] UNSCEAR (United Nations Scientific Committee on the Effects of Atomic Radiation). (2016). Sources, Effects and Risks of Ionizing Radiation, United Nations, New York. http://www.unscear.org/docs/publications/2016/UNSCEAR_2016_Annex-D.pdf.

[24] Sahoo, S. K., Jha, V. N., Patra A. C., Jha S. K., Kulkarni M. S. Scientific background and methodology adopted on derivation of regulatory limit for uranium in drinking water – A global perspective. *Environmental Advances*, 2020, *2*(100020). https://doi.org/10.1016/j.envadv.2020.100020.

[25] IAEA (2018), IAEA-TECDOC-1842, Geological Classification of Uranium Deposits and Description of Selected Examples. *International Atomic Energy Agency,* IAEA TECDOC series, IAEAL 18-01150, ISBN 978–92–0–101618–8. https://www-pub.iaea.org/MTCD/Publications/PDF/TE1842_web.pdf.

[26] Samaropoulos, I., Efstathiou, M., Pashalidis, I., Ioannidou, A. Determination of uranium concentration in ground water samples of Northern Greece. In *EPJ Web of Conferences*. 2012, *24*, 03005. EDP Sciences.

[27] Nriagu, J., Nam, D. H., Ayanwola, T. A., Dinh, H., Erdenechimeg, E., Ochir, C., Bolormaa, T. A. High levels of uranium in groundwater of Ulaanbaatar, Mongolia. *Science of the Total Environment*, 2012, *414*, 722-726. https://doi.org/10.1016/j.scitotenv.2011.11.037.

[28] Eggers, M. J., Moore-Nall, A. L., Doyle, J. T., Lefthand, M. J., Young, S. L., Bends, A. L., Crow Environmental Health Steering Committee; Camper, A. K. Potential Health Risks from Uranium in Home Well Water: An Investigation by the Apsaalooke (Crow) Tribal Research Group. *Geosciences*, 2015, *5*(1), 67-94. https://doi.org/10.3390/geosciences5010067.

[29] Kopylova, Y., Guseva, N., Shestakova, A., Khvaschevskaya, A., Arakchaa, K. Uranium and thorium behavior in groundwater of the natural spa area "Choygan mineral water" (East Tuva). In *IOP conference series: Earth and environmental science,* 2015, *27*(1), 012034. IOP Publishing.

[30] ICRP. (2008). Recommendations of the International Commission on Radiological Protection. Annals of the ICRP.

[31] Banning, A., Benfer, M. Drinking water uranium and potential health effects in the German Federal State of Bavaria. *International journal of environmental research and public health*, 2017, *14*(8), 927.

[32] Sahoo, S. K., Jha, S. K., Jha, V. N., Patra, A. C., Kulkarni, M. S. Survey of uranium in drinking water sources in India: interim observations. *Current Science*, 2021, 120(9), 1482-1490. https://doi.org/10.18520/cs/v120/i9/1482-1490.

[33] Bansal, V., Tyagi, R. K., Prasad, R. Determination of uranium concentration in domestic water samples by fission track method. *Journal of Radioanalytical and Nuclear Chemistry Articles*, 1988, *125*(2), 439–443. https://doi.org/10.1007/BF02041701.

[34] Ramola, R. C., Singh, S., Virk, H. S. Uranium and radon estimation in some water samples from Himalayas. *International Journal of Radiation Applications and Instrumentation. Part D. Nuclear Tracks and Radiation Measurements*, 1988, *15*(1-4), 791-793. https://doi.org/10.1016/1359-0189(88)90252-X.

[35] Singh, P., Rana, N. P. S., Azam, A., Naqvi, A. H., Srivastava, D. S. Levels of uranium in waters from some Indian cities determined by fission track analysis. *Radiation Measurements*, *26*(5), 1996, 683–687. https://doi.org/10.1016/S1350-4487(97)82882-X.

[36] Singh, B., Singh, G., Sandhu, A. S., Singh, S. Uranium estimation in water samples collected from some areas of Himachal Pradesh, India. Radiation measurements, 1999, *31*(1-6), 683-685.

[37] Singh, S., Malhotra, R., Kumar, J., Singh, B., Singh, L. (2001). Uranium analysis of geological samples, water and plants from Kulu area, Himachal Pradesh, India. *Radiation Measurements*, 2001, *34*(1–6), 427–431. https://doi.org/10.1016/S1350-4487(01)00200-1.

[38] Singh, S., Rani, A., Mahajan, R. K., Walia, T. P. S. Analysis of uranium and its correlation with some physico-chemical properties of drinking water samples from Amritsar, Punjab. *Journal of Environmental Monitoring*, 2003, *5*(6), 917–921. https://doi.org/10.1039/b309493f.

[39] Mehra, R., Singh, S., Singh, K. Uranium studies in water samples belonging to Malwa region of Punjab, using track etching technique. *Radiation Measurements*, 2007, *42*(3), 441–445. https://doi.org/10.1016/j.radmeas.2007.01.040.

[40] Babu, M. N. S., Somashekar, R. K., Kumar, S. A., Shivanna, K., Krishnamurthy, V., Eappen, K. P. Concentration of uranium levels in groundwater. *International Journal of Environmental Science and Technology*, 2008, *5*(2), 263–266. https://doi.org/10.1007/BF03326020.

[41] Kansal, A., Siddiqui, N. A., Gautam, A. Assessment of heavy metals and their interrelationships with some physicochemical parameters in eco-efficient rivers of Himalayan region. *Environmental Monitoring and Assessment*, 2013, *185*, 2553-2563. https://doi.org/10.1007/s10661-012-2730-x.

[42] Balbudhe, A. Y., Srivastava, S. K., Vishwaprasad, K., Srivastava, G. K., Tripathi, R. M., Puranik, V. D. Assessment of age-dependent uranium intake due to drinking water in Hyderabad, India. *Radiation Protection Dosimetry,* 2012, *148*(4), 502–506. https://doi.org/10.1093/rpd/ncr193.

[43] Byju, S. B., Sunil, A., Reeba, M. J., Christa, E. P., Vaidyan, V. K., Prasad, R., Jojo, P. J. Uranium in drinking water from the south coast districts of Kerala, India. *Iranian Journal of Radiation Research*, *10*(1), 2012, 31–36.

[44] Bhangare, R., Tiwari, M., Ajmal, P., Sahu, S., Pandit, G. Laser flourimetric analysis of uranium in water from Vishakhapatnam and estimation of health risk. *Radiation Protection and Environment*, 2013, *36*(3), 128–132. https://doi.org/10.4103/0972-0464.137478.

[45] Rani, A., Mehra, R., Duggal, V., Balaram, V. Analysis of uranium concentration in drinking water samples using ICPMS. *Health Physics*, 2013, *104*(3), 251–255. https://doi.org/10.1097/HP.0b013e318279ba05.

[46] Singh, B., Garg, V. K., Yadav, P., Kishore, N., Pulhani, V. Uranium in groundwater from Western Haryana, India. *Journal of Radioanalytical and Nuclear Chemistry*, 2014, *301*(2), 427–433. https://doi.org/10.1007/s10967-014-3133-y.

[47] Jakhu, R., Mehra, R., Mittal, H. M. Exposure assessment of natural uranium from drinking water. *Environmental Science: Processes and Impacts*, 2016, *18*, 1540–1549. https://doi.org/10.1039/c6em00514d.

[48] Selvi, B. S., Vijayakumar, B., Rana, B. K., Ravi, P. M. Distribution of natural uranium in groundwater around Kudankulam. *Radiation Protection and Environment*, 2016, *39*(1), 25–29. https://doi.org/10.4103/0972-0464.185164.

[49] Bajwa, B. S., Kumar, S., Singh, S., Sahoo, S. K., Tripathi, R. M. Uranium and other heavy toxic elements distribution in the drinking water samples of SW-Punjab, India. *Journal of Radiation Research and Applied Sciences*, 2017, *10*(1), 13–19. https://doi.org/10.1016/j.jrras.2015.01.002.

[50] Sar, S. K., Diwan, V., Biswas, S., Singh, S., Sahu, M., Jindal, M. K., Arora, A. Study of uranium level in groundwater of Balod district of Chhattisgarh state, India and assessment of health risk. *Human and Ecological Risk Assessment*, 2017, *24*(3), 691–698. https://doi.org/10.1080/10807039.2017.1397498.

[51] Sar, S. K., Sahu, M., Singh, S., Diwan, V., Jindal, M. K., Arora, A. Assessment of uranium in ground water from Durg District of Chhattisgarh state and its correlation

with other quality parameters. *J Radioanal Nucl Chem*, 2017, *314*, 2339–2348. https://doi.org/10.1007/s10967-017-5587-1.

[52] Chandramohan, J., Ganesh, D., Harikrishnan, N., Ravisankar, R., Chandrasekaran, A., Neelakantan, M. A., Raja, V. Study of Uranium Concentration in Drinking Water in Tiruvannamalai District, Tamil Nadu, India and Assessment of Health Risk. *Journal of Radiation and Nuclear Applications*, 2018, *3*(2), 103–110. https://doi.org/10.18576/jrna/030205.

[53] Narang, S., Kumar, D., Kumar, A. Health-related analysis of uranium in Fazilka district, Punjab, India. *Current Science*, 2018, *115*(11), 2079–2084. https://doi.org/10.18520/cs/v115/i11/2079-2084.

[54] Sharma, T., Sharma, A., Kaur, I., Mahajan, R. K., Litoria, P. K., Sahoo, S. K., Bajwa, B. S. Uranium distribution in groundwater and assessment of age dependent radiation dose in Amritsar, Gurdaspur and Pathankot districts of Punjab, India. *Chemosphere*, 2019, *219*, 607–616. https://doi.org/10.1016/j.chemosphere.2018.12.039.

[55] Naik, M. R., Barik, M., Jha, V. N., Sahoo, S. K., Sahoo, N. K. Access of Ground Water Quality with Associated Physico-Chemical Parameters of Cuttack District, Odisha, India. *International Journal of Innovative Technology and Exploring Engineering*, 2019, *8*(12), 260–267. https://doi.org/10.35940/ijitee.L1072.10812S19.

[56] Sahu, M., Sar, S. K., Baghel, T., Dewangan, R. Seasonal and geochemical variation of uranium and major ions in groundwater at Kanker district of Chhattisgarh, central India. *Groundwater for Sustainable Development*, 2020, *10*, 100330. https://doi.org/10.1016/j.gsd.2020.100330.

[57] Kale, A., Bandela, N., & Kulkarni, J. Assessment of chemo-radiological risk of naturally occurred uranium in groundwater from the Beed district, India. *Journal of Radioanalytical and Nuclear Chemistry*, 2020, *323*(1), 151–157. https://doi.org/10.1007/s10967-019-06849-3.

[58] Das, A., Das, S. S., Chowdhury, N. R., Joardar, M., Ghosh, B., Roychowdhury, T. Quality and health risk evaluation for groundwater in Nadia district, West Bengal: An approach on its suitability for drinking and domestic purpose. *Groundwater for Sustainable Development*, 2020, *10*, 100351. https://doi.org/10.1016/j.gsd.2020.100351.

[59] Shashi, P. S., Bholey, S., Singh, B. M., Gautam, Y. P., Balaram, P. Geochemical signatures and chemical toxicity assessment of groundwater uranium in the agricultural dominated tehsil after indian summer monsoon. *Research Journal of Chemistry and Environment*, 2020, *24*(10), 75–82.

[60] Singh, S. P., Dutta, P. K. Geospatial Groundwater Quality Assessment and Identification of Polluted Risky Regions in Jamui District of Bihar with Special Reference to Uranium and Fluoride Concentration. *Current World Environment*, 2020, *15*(3), 653–662. https://doi.org/10.12944/cwe.15.3.30.

[61] Kumar, A., Arora, T., Singh, P., Singh, K., Singh, D., Pathak, P. P., Ramola, R. C. Quantification of radiological dose and chemical toxicity due to radon and uranium in drinking water in Bageshwar region of Indian Himalaya. *Groundwater for*

Sustainable Development, 2021, *12*, 100491. https://doi.org/10.1016/j.gsd.2020.100491.

[62] Saikia, R., Chetia, D., Bhattacharyya, K. G. Estimation of uranium in groundwater and assessment of age-dependent radiation dose in Nalbari district of Assam, India. *SN Applied Sciences*, 2021, *3*(1), 1–12. https://doi.org/10.1007/s42452-020-04071-5.

[63] Lale, K., Prakash, D., Srivastava, S. K., Mohiddin, S. K., Nagar, A. Uranium Contamination in Shallow Aquifers of Haryana State, India. *Int. Res. Jour. Eng. Tech.,* 2021, *8*, 3364-3372.

[64] Shalumon, C. S., Sanu, K. S., Thomas, J. R., Aravind, U. K., Radhakrishnan, S., Sahoo, S. K., Jha, S. K., Aravindakumar, C. T. Analysis of uranium and other water quality parameters in drinking water sources of 5 districts of Kerala in southern India and potability estimation using water quality indexing method. *HydroResearch*, 2021, *4*, 38–46. https://doi.org/10.1016/j.hydres.2021.04.003.

[65] Patni, K., Pande, C., Pande, A. P., Tewari, G., Joshi, T. Distribution of naturally occurring uranium and other heavy toxic elements in selected spring water samples of Pithoragarh District, Uttarakhand, India. *SN Applied Sciences*, 2020, *2*(12), 2129. https://doi.org/10.1007/s42452-020-03947-w.

[66] Patni, K., Pande, C., Pande, A. P., Tiwari, G., Joshi, T. Seasonal variation of uranium and physico-chemical parameters in spring water sources of Pithoragarh city, Uttarakhand, India. *Journal of Radioanalytical and Nuclear Chemistry*, 2021, *329*(2), 647–660. https://doi.org/10.1007/s10967-021-07823-8.

[67] Srinivasan, R., Pandit, S. A., Karunakara, N., Salim, D., Kumara S. K., Kumar M. R., Khatei G., Ramkumar K. D. High uranium concentration in groundwater used for drinking in parts of eastern Karnataka, India. *Current Science*, 2020, 121(11), 1459-1469.

[68] Yadav, A. K., Sahoo, S. K., Mahapatra, S., Kumar, A. V., Pandey, G., Lenka, P., Tripathi, R. Concentrations of uranium in drinking water and cumulative, age-dependent radiation doses in four districts of Uttar Pradesh, India. *Toxicological and Environmental Chemistry*. 2014, *96*(2), 192-200. https://doi.org/10.1080/02772248.2014.934247.

[69] Pant, D., Keesari, T., Sharma, D., Rishi, M., Singh, G., Jaryal, A., Sinha, U. K., Dash, A., Tripathi, R. M. Study on uranium contamination in groundwater of Faridkot and Muktsar districts of Punjab using stable isotopes of water. *Journal of Radioanalytical and Nuclear Chemistry*, 2017, *313*, 635-639. https://doi.org/10.1007/s10967-017-5284-0.

[70] World Health Organization. (2001). Depleted uranium: sources, exposure and health effects. *World Health Organization*. https://apps.who.int/iris/handle/10665/66930.

[71] Foo, K. Y., Hameed, B. H. An overview of dye removal via activated carbon adsorption process. *Desalination and Water Treatment*, 2010, *19*(1–3), 255–274. https://doi.org/10.5004/dwt.2010.1214.

[72] Foo, K. Y., Hameed, B. H. Potential of activated carbon adsorption processes for the remediation of nuclear effluents: A recent literature. *Desalination and Water Treatment*, 2012, *41*(1–3), 72–78. https://doi.org/10.1080/19443994.2012.664680.

[73] Katsoyiannis, I. A., Zouboulis, A. I. Removal of uranium from contaminated drinking water: a mini review of available treatment methods. *Desalination and Water Treatment*, 2013, *51*:13-15, 2915-2925. https://doi.org/10.1080/19443994.2012.748300.

[74] Kausar A., Bhatti H. N. Adsorptive Removal of Uranium from Wastewater: A Review. *Journal- Chemical Society of Pakistan,* 2013, *35*(3), 041-1052.

[75] Rosenberg, E., Pinson, G., Tsosie, R., Tutu, H., Cukrowska, E. *Uranium Remediation by Ion Exchange and Sorption Methods: A Critical Review Various types of solid phase sorbents are studied and evaluated.* 2016, *60*(1), 59–77.

[76] Chiarizia, R., Horwitz, E. P. Study of uranium removal from groundwater by supported liquid membranes. *Solvent Extraction and Ion Exchange*, 1990, *8*(1), 65–98. https://doi.org/10.1080/07366299008917987.

Chapter 6

Research Vignettes of Transport, Speciation, and Kinetics of Uranium from Mine Wastes

José M. Cerrato*
Sumant Avasarala
Carmen A. Velasco
and Maria I. Meza
Department of Civil, Construction & Environmental Engineering,
University of New Mexico, Albuquerque, New Mexico, USA
Center for Water and the Environment, Albuquerque, New Mexico, USA

Abstract

Within this chapter, we summarize recent studies relevant to uranium fate and transport in the lands surrounding uranium mining activities, focusing on tribal lands in the Southwestern United States. The findings, however, will be applicable in general to issues of uranium transport in water, especially in geologically similar regions. The coincident transport of uranium, vanadium, and arsenic are detailed based on physical and chemical interactions. Studies of these heavy metal interactions with natural dissolved organic matter and sorption processes in the environment are also described.

* Corresponding Author's Email: jcerrato@unm.edu

In: The Chemistry of Uranium
Editor: William A. Alexander
ISBN: 979-8-89530-802-8

1. Introduction

1.1. Legacy of Uranium Mining in the US

Uranium mining operations occurred during the 1900s to meet the federal government's demand for nuclear weapons after the Manhattan project [1, 2] and, to a more limited extent, for nuclear energy generation [3, 4]. These mining activities left a legacy of thousands of abandoned mines all over the United States [5] and have resulted in numerous sites having elevated metal concentrations in water and soil [6-8]. The western part of the United States has a legacy of over 160,000 abandoned mines [5] and represents a particularly extensive area of exposure to mine wastes. Uranium (U), and other metals concentrations in drinking water sources exceed United States Environmental Protection Agency (EPA) and World Health Organization (WHO) standards in the U.S. [6, 9-11] The problem is particularly important in regions where U mining has occurred. An as example of the scale of the problem, in the western US approximately 15,000 abandoned uranium mine waste sites have been identified, the majority of which are located in Colorado, Utah, New Mexico, Arizona, and Wyoming [5, 12].

Following the Second World War (WWII), many Federal and Native American lands were intensively mined for uranium to meet the increasing demands for nuclear weapons. Seventy-five percent of uranium mine sites are on federal and tribal lands [5, 12]. Several Native American tribes or pueblos in the southwestern US have had large U mines located on their lands. To exemplify the national problem, about 1100 abandoned uranium mine waste sites located on the Navajo Nation have been overlooked for decades [10, 11, 13]. Between 1944 and 1986, nearly 30 million tons of uranium ore were extracted from land that was part of the Navajo Nation [10, 11, 13], leaving significant quantities of post-extraction waste materials. Improper disposal and handling of these wastes have resulted in many sites across the United States (US) having contaminated surrounding land, water (groundwater and surface water), and air resources. To propose appropriate remediation strategies, there is a need to understand the mechanisms that affect the transport of U and other co-occurring metals from mine wastes under relevant environmental conditions.

1.2. Risks to Environment and Human Health

Mining operations generate two types of waste: mine waste and mill tailings. Mine wastes are unregulated, higher concentration U ores that have been abandoned after mining operations cease, while the mill tailings are federally-regulated, low concentration wastes left after processing the ore for U [14, 15]. We focus in this chapter on mine wastes which are generated during mining operations, as these unregulated wastes potentially pose a larger risk to human and environmental health and safety. Mine wastes contain mixtures of uranium (U), vanadium (V), arsenic (As), and other co-occurring metals [10, 13, 16]. These wastes pose environmental and health risks to nearby communities. However, before effective remediation approaches can be proposed, we need to know relevant details about the sites and the distribution means for the waste.

Past and on-going studies have documented the impact of these materials on the waters of the region. Studies have reported U concentrations of 35.3-772 $\mu g\ L^{-1}$ in surface waters near the Pueblo of Laguna, New Mexico (NM), neighboring the Jackpile Mine, one the largest uranium mines in the US [6]. These concentrations exceed the US Environmental Protection Agency maximum contaminant level of 30 $\mu g\ L^{-1}$ [9]. Additionally, the U concentration detected in solid samples from the Jackpile Mine, adjacent to the Rio Paguate, was 9300 $mg\ kg^{-1}$ [6]. The Rio Paguate flows from Miocene basalt and andesite flows of Mount Taylor, southeast towards the village of Paguate through the Jurassic Morrison Formation and the abandoned Jackpile-Paguate Mine [17]. Reactions between these surface waters and U mine wastes could affect water quality and pose a potential human risk.

The Jackpile Mine has been listed in the EPA National Priorities List for the Comprehensive Environmental Response, Compensation, and Liability Act (CERCLA) cleanup [18-20]. Uranium is regulated because it can cause cancer, kidney toxicity, miscarriages, birth defects, and heart disease [21-24]. Recent evidence suggests that residential proximity to abandoned mine sites results in increased serum cumulative inflammatory potential, independent of metals-intake via drinking water [23]. A study in the Grants Mining District located in western NM reports that U-bearing dust affects the extent of U dissolution in simulated lung fluids, which could have potential health implications [25].

Metal concentrations that may impact human health for Native American people were also found to be correlated to mine source distance [23]. Elevated concentrations of U and V in water resources pose a concern for health risks to individuals in the surrounding communities [21-27]. For example, the adverse toxic effects of U and V are well documented in human exposures by inhalation and by ingestion [28-33].

The co-occurrence of U and As caused by natural and anthropogenic processes is a concern to communities living in the proximity of contaminated sites [34-40]. The concern of course is not limited to the SW US as minerals containing U and As occur in various locations worldwide [40-42]. For example, beyond the several Southwestern US sites we predominately focus on herein, some of the largest U mines are located in Key Lake, Saskatchewan, Canada and the St. Austell District of Cornwall, England [40-42]. Communities that surround the mine sites have been impacted by the release of U and As, which is of major concern for public health as both are toxic and carcinogenic, and U is radioactive [34, 43-44].

Uranium exposure has long been known to cause cancer (e.g., bone, reproductive, and gastric), and other adverse human health outcomes [25, 31, 45-48], but has also been linked to adverse developmental outcomes in animals [49-50]. Additionally, recent studies in Navajo communities have linked exposures to mine wastes with an increased likelihood for developing one or more chronic diseases including hypertension, kidney disease, and diabetes [51].

Human health is not the only concern as animal and plant life are also affected by mine wastes. For instance, these dissolved metals have the ability to act as Lewis acids, complexing with the hydroxyl group (OH^-) and causing a drop in pH. Acidic conditions in these sites can also be caused due to acid mine drainage, which is the outflow of acidic water from a mining site caused primarily from oxidation of iron sulfide (FeS_2) or pyrite [52-53]. An acidic environment is generally toxic to biota and less favorable for many other organisms than is an environment at circumneutral pH conditions [54].

2. Physical and Chemical Interactions in Uranium and Vanadium Transport from Mine Wastes

Within this section, we detail a series of research studies that provide insights on the mechanisms that affect the reactive transport of U and V from abandoned uranium mine wastes. Information on the reactivity and mobility of U and V can be key to understanding the potential metal exposure pathways in abandoned uranium mine waste sites.

After a review of relevant background literature, we present within these vignettes a summary of our work on:

- Vignette A) The reactive transport of U and V from the Blue Gap/Tachee Claim-28 mine waste site. We see that dissolution of the U-V bearing mineral is a relevant mechanism that affects the reactive transport of U and V from the Blue Gap/Tachee Claim-28 abandoned U mine.
- Vignette B) The crystal chemistry of U-V bearing minerals occurring in the Blue Gap/Tachee Claim-28 mine wastes. U-V bearing minerals in the Blue Gap/Tachee mine wastes were identified as hydrated carnotites that contain carbon inclusions.

2.1. Background

Mine waste sites located on Native American lands have been overlooked due to low population densities and site remoteness. The Blue Gap/Tachee Claim-28 mine site in northeastern Arizona and the Jackpile-Paguate mine waste site in Laguna Pueblo, New Mexico, are two such sites that are indicative of this situation.

The Blue Gap/Tachee Chapter of the central Navajo Nation, located approximately 40 miles west of Chinle, AZ, started operating during the early 1950s. On the mesa tops and steep slopes of this location, 13 different mines operated from 1954-1968 [14]. In the early 1990s, initial reclamation efforts agglomerated the wastes in the Claim 28 site under a layer of uncontaminated soil (SMCRA) [20, 55]. At Blue Gap/Tachee, the uranium ores occur in tabular and lenticular sandstone units of the Rough Rock Sandstone member in the Toreva formation. A recent study conducted by our research group in the Claim 28 site identified the major mineralogical constituents of this arkosic sandstone to be 59% quartz (SiO_2), 34% potassium feldspar ($KAlSi_3O_8$), and 7% kaolinite ($Al_2Si_2O_5(OH)_4$) [14, 55-56]. The dissolution of U-V bearing minerals from mine wastes that were potentially affect the elevated concentrations (with respect to the maximum contaminant level [MCL] for U) in neighboring water resources (67−169 μg/L of U) at the site [14]. U-V bearing minerals from the site were found to have a structure similar to that of carnotite ($K_2(UO_2)_2(VO_4)_2{\cdot}3H_2O$) or tyuyamunite ($Ca(UO_2)_2V_2O_8{\cdot}(5\text{-}8)H_2O$) [14]. However, there is no information on the type of uranyl vanadates these

U-V bearing minerals represent or the processes that affect U-mobility from these mine wastes. This information is critical to better understand mobility of U and V from many U mines around the world where uranyl vanadates are predominant. For instance, abandoned uranium mines in the Colorado Plateau, South Dakota, southwest China, southern Jordan, Korea and the calcreted drainages of arid and semiarid western and southern Australia are rich in carnotite, tyuyamunite, and other U-V bearing minerals that have affected the water quality of neighboring surface and ground water resources [57-64].

The Jackpile-Paguate mine is an abandoned uranium mine located on the 1093 Laguna reservation in north-western New Mexico [65], 45 miles west of Albuquerque, NM, USA. Approximately 600 m away from the Jackpile-Paguate Mine is Paguate, one among the eight major villages of Laguna Pueblo. Just below Paguate lies one of the most extensive uranium deposits in the US. The mining operation at the Jackpile-Paguate mine disturbed approximately 2,656 acres of property [66] and was once the largest operational, open pit uranium mine in the world. The U deposits in the Jackpile-Paguate Mine are present in the Jackpile sandstone found in the uppermost part of the Brushy Basin Shale Member of the Jurassic and Cretaceous strata [67-69]. Although the combined ores in the Jackpile-Paguate mine are 2 km long and 600 m wide, the semi-tabular U deposits in the Jackpile sandstone are only about ~5 m thick [67-69].

As mentioned, the Rio Paguate flows the abandoned Jackpile-Paguate Mine [17], potentially affecting water quality posing a human health risk. For instance, a study by our research group reported seasonal variations in elevated concentrations of U and other co-occurring metals in Rio Paguate (35.3 to 772 mg L^{-1}) that were later identified as a consequence of dissolution of U-bearing minerals from these U-ore deposits [6]. However, the geochemical processes that affect U mobility from these mine wastes is still not well understood.

2.1. State of Knowledge About Processes Affecting U and V Transport from Mine Wastes

The mobility of U and V are affected by a variety of biogeochemical and physical processes. Uranium is an actinide abundantly found in various parts of the southwestern US with alternating valence states among IV, V, and VI depending on the surface and subsurface conditions in which it is present [55-58]. The +6 oxidation state (U(VI)) is the most mobile and toxic form of U [28-30]. The processes that affect U mobility are driven by physicochemical parameters, including pH, temperature, reaction kinetics, and elemental composition of the system [70-74]. Uranium mobility is enhanced by the presence of complexants like carbonate and bicarbonate that form aqueous complexes like U-Ca-CO_3^{-2} and U-CO_3^{-2} [22, 75]. Combined use of these natural complexants has been reported to be effective for U mobility in systems with high calcium (Ca), sulfur (S), manganese (Mn), and aluminum (Al) concentrations [52]. Mining industries use these complexants in alkaline leaching for targeted U extraction [76].

The aqueous complexes formed from such complexation reactions adsorb onto the oppositely charged surfaces of SiO_2, hydrous ferric oxide, ferrihydrite, aluminum hydroxides, clays, and organic matter [77-83]. Uranyl complexes bind strongly onto organic matter through either ion-exchange, coordination/complexation, donor-receptor interactions, or van der Waals forces at circumneutral conditions, substantially enhancing U mobility [83-88]. The degree to which U mobility is enhanced depends on 1) the electrostatic potential between aqueous complexes and organic matter surfaces, 2) hydrophobicity of organic matter, and 3) the ability of U(VI)-organic matter colloids to stay suspended [88-90]. Like aqueous U(VI), monomeric U(IV), which represents non-crystalline U(IV) species, also adsorb onto organic matter aggregates and organic-carbon coated clays that possess higher mobility and toxicity considering their susceptibility to desorb and reoxidize [82, 91]. Only recently were monomeric biogenic U(IV) species found to be present at undisturbed roll front ore deposits [92]. In addition to adsorption, the tetravalent and hexavalent U also form bidentate aqueous complexes with organic matter [93].

Formation of sparingly soluble, biogenic, and chemogenic U(IV) through reduction reactions has been used as a technique to immobilize aqueous U [94-96]. The reduction of aqueous U(VI) can result in the formation of crystalline, submicron, biogenic, and chemogenic U(IV) phases whose dissolution rates and energetics are very similar to those of coarser U(IV) phases [97-99]. Submicron U particles and U adsorbed onto colloids of organic matter and oxides form about 30-40% of the total naturally occurring mobile U [100, 101]. In addition to sorption, U mobility can also be affected by dissolution/precipitation of U minerals. For instance, U occurs in the environment as vanadate, silicate, phosphate, carbonate, and oxide minerals [102-109].

Similarly, V is a redox-active transition metal that co-occurs with U at oxidation states from II to V [4, 110-113]. Vanadium is most soluble as V(IV), and V(V) [110-113], among which V(V) is considered most mobile and toxic

[28-30]. Like U, V is also mobilized by carbonate reagents [14]. However, unlike U, it does not complex with dissolved carbonate. Instead, its oxyanionic behavior causes the formation of aqueous anionic [$V_{10}O_{26}(OH)_2^{4-}$, $V_{10}O_{27}(OH)^{5-}$, $V_{10}O_{28}^{6-}$, $VO_2(OH)^{2-}$, $VO_3(OH)^{2-}$, VO_4^{3-}, $V_2O_6(OH)^{3-}$, $V_2O_7^{4-}$, $V_3O_9^{3-}$ and $V_4O_{12}^{4-}$], cationic [VO^{2+}] and neutral [$VO(OH)_3$] species [71, 73, 114-115]. These aqueous complexes of U and V adsorb onto ferrihydrite, aluminum hydroxides, and clays through electrostatic interactions [60, 64, 67-69, 73, 74, 92-102, 115-117], depending on the pH_{pzc} of the solids. U and V could also be immobilized through precipitation of secondary mineral phases such as carnotite ($K_2(UO_2)_2(VO_4)_2 \cdot 3H_2O$ and vanadinite ($Pb_5(VO_4)_3Cl$) [57, 74, 109]. These geochemical processes are key to the aqueous and solid speciation of U and V.

While previous studies have identified some factors that affect transport of U or V individually, specific mechanisms affecting the transport of U and other co-occurring metals in abandoned mine waste sites remains poorly understood. For instance, in a recent study the mobilities of both U and arsenic (As) from a U mill tailing site in South Dakota were observed [118]. Yet the specific processes affecting the transport of these metals are unknown [104]. Previous investigations have provided limited information on the release of metal mixtures from uranium mine wastes, and the interactions, co-occurrence, and behavior of these metal mixtures in the environment under relevant field conditions.

2.2 Vignette A: Reactive Transport of U and V from Abandoned Uranium Mine Wastes

Through this first vignette, we obtain a better understanding of processes affecting the transport of U and V from abandoned mine wastes collected from the Blue Gap/Tachee Claim-28 mine, AZ and Jackpile-Paguate mine, Laguna, NM. The Blue Gap/Tachee Claim-28 mine wastes were investigated by integrating flow-through column experiments with reactive transport modeling and electron microscopy. The samples were sequentially reacted in flow-through columns at pH 7.9 (10 mM $NaHCO_3$) and pH 3.4 (10 mM CH_3COOH) to evaluate the effect of environmentally relevant conditions encountered at Blue Gap/Tachee on the release of U and V. The reaction rate constants (k_m) for the dissolution of uranyl-vanadate (U-V) minerals predominant at Blue Gap/Tachee were obtained from simulations with the reactive transport software, PFLOTRAN [119-121]. The estimated reaction rate constants were within one order of magnitude for pH 7.9 ($k_m = 4.8 \times 10^{-13}$ mol cm^{-2} sec^{-1}) and pH 3.4 ($k_m = 3.2 \times 10^{-13}$ mol cm^{-2} sec^{-1}). However, the estimated equilibrium constants (K_{eq}) for U-V bearing minerals were more than six orders of magnitude different for reaction at circumneutral pH ($K_{eq} = 10^{-38.65}$) compared to acidic pH ($K_{eq} = 10^{-44.81}$). These results coupled with electron microscopy data suggest that the release of U and V is affected by water pH and the crystalline structure of U-V bearing minerals.

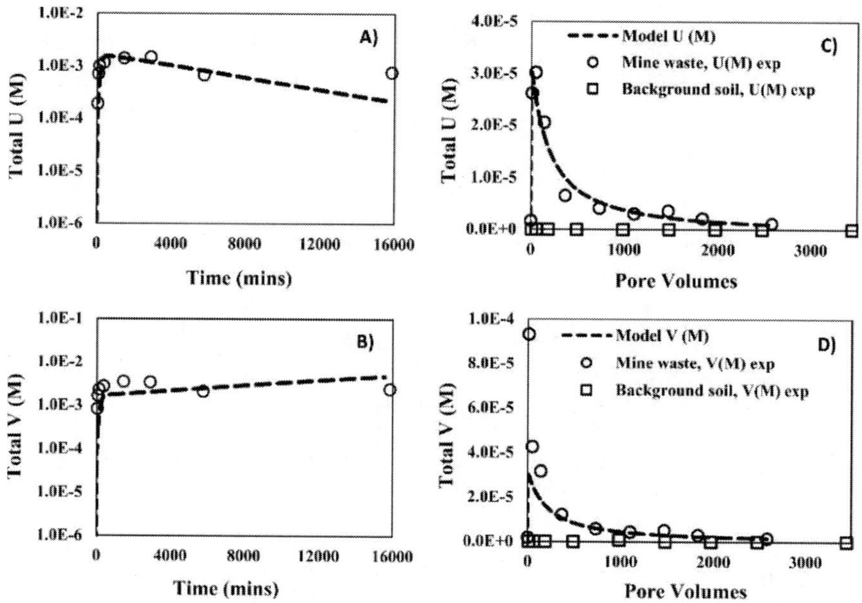

Figure 1. Measured and simulated effluent concentrations and reactive transport model (PFLOTRAN) of U and V, from mine waste (circle) and background soil (squares) during batch and continuous flow-through column experiments at circumneutral pH (using 10 mM HCO^{3-}). (A) U concentrations from batch experiments versus time; (B) V concentration from batch experiments versus time; (C) U concentrations from column experiments versus pore volumes; (D) V concentration from column experiments versus pore volumes. The curve fitting resulting from the reactive transport model are presented with dashed lines. Adapted with permission from Ref. [109]. Copyright 2014 American Chemical Society.

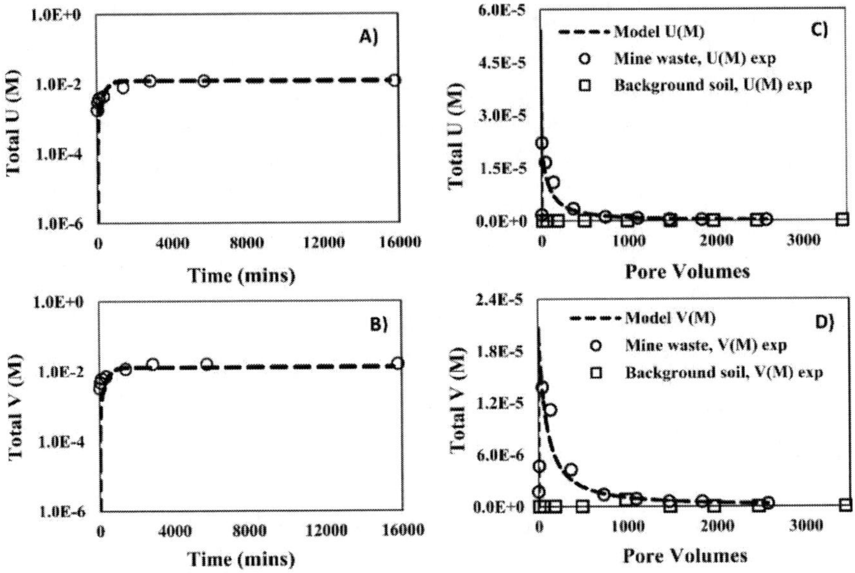

Figure 2. Measured and simulated effluent concentrations and reactive transport model (PFLOTRAN) of U and V, from mine waste (circle) and background soil (squares) during batch and continuous flow-through column experiments at acidic pH (using 10 mM $C_6H_8O_6$ and CH_3COOH). (A) U concentrations from batch experiments versus time; (B) V concentration from batch experiments versus time; (C) U concentrations from column experiments versus pore volumes; (D) V concentration from column experiments versus pore volumes. The curve fittings resulting from the reactive transport model are presented with dashed lines. Adapted with permission from Ref. [109]. Copyright 2014 American Chemical Society

This work integrates reactive transport modeling, electron microscopy, and aqueous chemistry methods to evaluate the mechanisms affecting the transport of U and V from mine wastes from a site in northeastern Arizona. Reactive transport modeling can be a useful tool to investigate the interfacial mechanisms that affect the surface and subsurface mobilization of U and other metals. Previous studies have used reactive transport modeling to investigate the adsorption behavior of U(VI) complexes onto alluvial aquifer sediments [119-120]. One outcome from this investigation includes the determination of solubility and reaction rate constants for the dissolution of U-V bearing mineral phases at circumneutral and acidic pH that can serve as a foundation to better understand their reactivity at relevant field conditions. This information can be useful to better understand the mobility of U and V in neighboring community water sources to assess risks for human exposure.

Additionally, the identification of factors affecting the dissolution of U-V bearing minerals under environmentally relevant conditions evaluated in this study is relevant to inform remediation and resource recovery initiatives in sites where these U-V bearing minerals are abundant. For instance, the results from this study have implications for other abandoned uranium mines in the Colorado Plateau, South Dakota (Black Hills), southwest China, southern Jordan, Korea, and the calcreted drainages of arid and semiarid western and southern Australia where carnotite ($K_2(UO_2)_2(VO_4)_2 \cdot 3H_2O$), tyuyamunite ($Ca(UO_2)_2V_2O_8 \cdot (5\text{-}8)H_2O$), and other U-V bearing minerals are commonly found [57-64].

Spectroscopy and microscopy analyses suggested that U is associated to different elements (see Figures 1 and 2), which is indicative of the complex mineralogy of these solids. These results are consistent with those reported in previous work [14]. Despite the chemical and mineralogical complexity of these mine wastes, dissolution of U-V bearing minerals was identified as the key mechanism controlling the reactive transport of U and V under the environmentally relevant conditions selected for this study. In addition, the integrated methodology used for this study is transferable to other mining and milling sites and to abandoned U mine waste sites where a better understanding of the reactive transport of U and its co-occurring elements is necessary.

2.3. Vignette B: Crystal Chemistry of Uranyl Vanadates in Mine Wastes

We investigated the crystal chemistry of U-V bearing minerals occurring in the Blue Gap/Tachee mine wastes using various spectroscopy, electron microscopy, and diffraction techniques. Specifically, samples were subjected to Raman Spectroscopy, X-ray Diffraction, Electron Microprobe Analysis, and Scanning Transmission Electron Microscopies. Raman Spectroscopy confirmed the carnotite structure of the uranyl vandate mine samples (Figure 3). The diffraction lines in the X-ray diffractogram (Figure 4) of these U-V bearing minerals showed agreement with most peaks of synthetic carnotite from the literature. The C in the U-V bearing mineral phase was found to co-occur as an organic carbon inclusion based on results from the bright field transmission electron microscopy (BFTEM) imaging (Figure 6B), energy filter transmission electron microscopy (EFTEM) (Figure 6C), and electron energy loss spectroscopy (EELS) (Figure 5). Quantitative electron microprobe analysis suggested the elemental composition of U-V bearing minerals was at a K:U:V ratio of 1:2:2.1 with an empirical stoichiometry similar to that of the anhydrous carnotite that differ in d-spacing. Discrepancies between measured (U-V bearing minerals) and calculated (anhydrous carnotite) diffraction ring patterns can be attributed to differences in the degree of hydration, suggesting the U-V bearing minerals were hydrated crystalline carnotite. Results from this investigation provide an improved understanding on the crystal chemistry of naturally occurring U-V bearing minerals in abandoned mine wastes.

Outcomes from this investigation suggest that the uranyl vanadate minerals in the Blue Gap/Tachee mine waste site were crystalline, elongated and platy grains (1-5 μm) of hydrated carnotites with significant intergranular porosity, rich in carbon inclusions (20-50 nm). These hydrated carnotites were susceptible to rapid amorphization on short-term exposure (<120s) to the electron beam, based on *d*-spacing information obtained using the electron diffraction capability of TEM. This information on the crystal chemistry of naturally occurring uranyl vanadate is critical to improve our understanding on their solubility that contribute to an improved understanding on the transport of U and V from these mine wastes. For instance, results from this investigation are coherent with our previous observations at Blue Gap/Tachee Claim-28 mine waste site, where the solubility of this hydrated carnotite (equilibrium constant: K_{eq} at circumneutral pH = $10^{-38.65}$ and acidic pH K_{eq} = $10^{-44.81}$) was similar to that of synthetic carnotite reported in literature (K_{eq} = $10^{-56.38}$) [109]. However, there were minor discrepancies in the solubility information that could be attributed to differences in crystal chemistry, highlighting the importance of crystal chemistry while understanding the reactive transport of metals. Knowledge obtained from this effort and the integrated methodology used to obtain this information can be applied to other mine waste sites where uranyl vanadates are abundant [57-64, 123-124].

3. Natural Organic Matter Interactions with Uranium Mine Wastes

The goal of the research presented as vignettes in this section was to understand the chemical interfacial reactions between U and natural organic matter (NOM) affecting the solubility of U and changes in the molecular composition of NOM at environmentally relevant pH. We begin with background on natural organic matter (NOM) chemistry, the occurrence of U and NOM in the environment, and their interactions with each other, before presenting the research vignettes. These vignettes summarize our work on:

Vignette C) Efforts toward better understanding the role of organic matter on U speciation under environmentally relevant pH conditions by investigating the functional group chemistry of natural organic matter (NOM) associated with U(IV) and U(VI) in solids from mineralized deposits from the Jackpile Mine. The main finding was that phenolic and carbonyl functional groups are more abundant in the organic matter fraction in these mineralized deposits in which both reduced U(IV) and oxidized U(VI) were present, which is contrary to previous (and sparse) other studies that saw higher fractions of carboxyl groups.

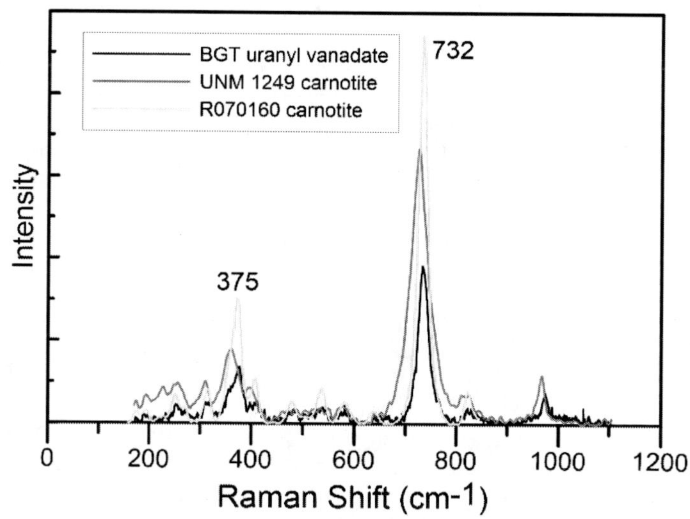

Figure 3. Raman spectrum for the BGT Claim 28 uranyl vanadate with Raman spectra for carnotite (R070160) from the Mineralogical Society of America RRUFF Raman spectra database and a carnotite specimen (1249) from the University of New Mexico Mineral Collection. Each spectrum is scaled differently for comparison. The major peaks at 732 and 375 cm^{-1} are indicated. There is very good agreement between the Raman spectra from BGT and R070160, with the major peak at 732 cm^{-1}, but the UNM carnotite shows a shift in the position of this peak to lower values. For comparison, the major peak for vandermeerscheite is at 745 cm^{-1}. Adapted from [122] under CC 4.0 license.

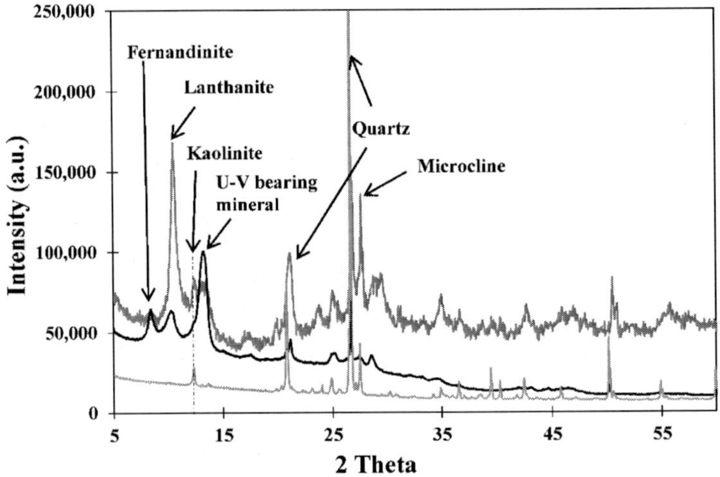

Figure 4. Powder x-ray diffraction pattern of Blue Gap Tachee mine waste. The bottom profile represents the diffractogram of the powdered mine waste as collected. The middle (darker) profile represents a diffractogram of the preferentially oriented mine waste sample obtained from the suspended fraction of a water separation technique with an overnight settling time. The top profile is a diffractogram of a randomly oriented mine waste sample obtained from the suspended fraction of a water separation technique with a settling time of <30 s. Adapted from [122] under CC 4.0 license.

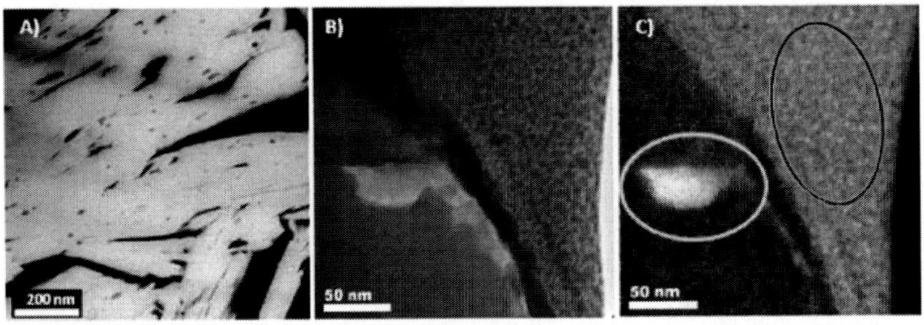

Figure 5. (A) HAADF images in scanning transmission electron microscopy (STEM) mode showing the co-occurrence of C (dark inclusions) within the uranyl vanadate from the Blue Gap Tachee Claim 28 mine waste. (B) Bright-field TEM image and (C) energy filtered transmission electron microscopy (EFTEM) image carbon map confirming that the inclusion in the uranyl vanadate is carbon-rich. The area in the white ellipse is a C-rich inclusion and the area in the black ellipse indicates carbon retained in the organometallic Pt compound that is used to protect the FIB section from ion beam damage during sample preparation. Adapted from [122] under CC 4.0 license.

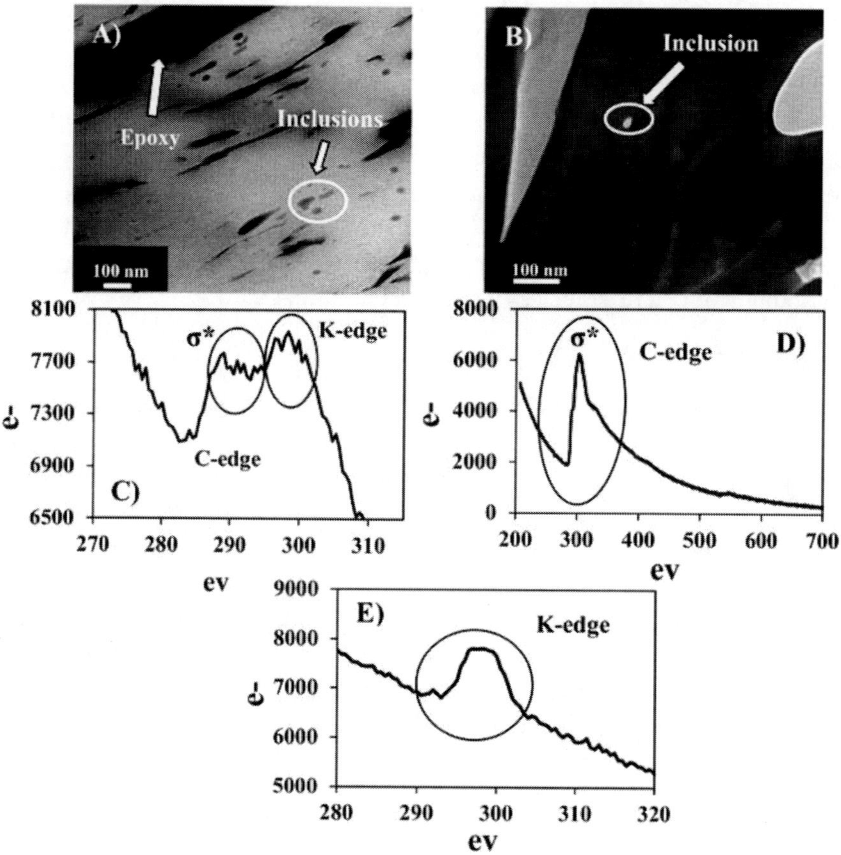

Figure 6. Electron energy loss spectroscopy (EELS) analysis of the carbon inclusions and epoxy. (A) HAADF scanning transmission electron microscopy (DF-STEM) image of the inclusions and epoxy (arrows) in the focused ion beam (FIB) section of the uranyl vanadate. (B) Bright-field (BF)-TEM image of an inclusion that was analyzed using EELS. (C) EELS spectrum of the C inclusion showing the presence of C and K-edge (circled areas). (D) EELS spectrum of the epoxy showing the presence of C-edge (circled). (E) EELS spectra of the uranyl vanadate away from the inclusion showing the presence of only K-edge (circled). The C L edge at 284 eV is missing demonstrating that the carbon is present only in the inclusions. Adapted from [122] under CC 4.0 license.

Vignette D) Investigating the effect of pH on the adsorption, precipitation, and solubilization reactions of U(VI) in the presence of NOM. This work highlights that adsorption of U onto particulate organic matter (POM) leads to the precipitation of U-bearing crystalline solids at pH 4 after 24 h. Results also showed that adsorption reactions are relevant as in the bulk, U(VI) was mostly adsorbed to POM at low pH. These results confirm that complexation of U and dissolved organic matter (DOM) favors the solubility of U at pH 7.

Vignette E) The effect of oxidizing conditions on co-occurring nano-particulate U(IV) and U(VI) phases in Jackpile mine wastes. The presence of submicron U(VI) and U(VI) phases were identfied in the organic rich sediments of the Jackpile-Paguate mine. The presence of coffinite and U-P-K bearing minerals were also identified in these mine wastes. The highest U release was observed during batch reactions with 10 mM $NaHCO_3$ solution containing ambient dissolved oxygen concentrations.

3.1. State of Knowledge of Natural Organic Matter Interactions

Natural organic matter (NOM) is a complex combination of heterogeneous organic molecules, resulting from the decomposition of animal and plant matter [22]. NOM consists of an extraordinarily complex mixture of compounds including carbohydrates, amino acids, and proteins [23]. Because of the molecular heterogeneity and complexity of NOM, it is often characterized based on its solubility. Natural organic matter is most abundant in the litter layers and upper mineral horizons of soils, and is transported to subsoils by water as dissolved molecules and fine suspended colloids or dissolved organic matter (DOM) [24, 125]. Dissolved organic matter is the organic matter fraction that passes through a 0.45-μm filter; while particulate organic matter (POM) refers to the organic particles remaining on a 0.45 μm filter [22, 126]. DOM affects biogeochemical processes, particle stability and transport, metal complexation, and production of disinfection byproducts (DBPs) during water and wastewater treatment [127-131].

The dissolved organic matter (DOM) fraction of NOM is closely associated with the mineral matrix of soil due to the strong complexation between surface metals and acidic organic ligands, particularly with those associated with aromatic structures [24]. Changes in pH and reactions with metals affect the molecular composition of DOM and POM.

Natural organic matter can also be operationally defined as a mixture of humic and nonhumic substances. Recognizable plant debris and organic compounds, such as polysaccharides, lignin, proteins, and polypeptides, are categorized as nonhumic substances [132]. The remaining high-molecular-weight, highly transformed, brown- to black-colored materials are considered humic substances (HS) [133]. Humic substances are often described as coiled, long-chain molecules or two- or three-dimensional cross-linked macromolecules with a wide range of organic functional groups. The molecular structure and chemistry of HS are still not well understood because of intrinsic chemical heterogeneity and geographical variability [134]. Thus, HS are also operationally fractioned based on their solubility [135]. Fulvic acid (FA) are soluble at all pH values, humic acids (HA) are insoluble at pH < 2 but soluble at higher pH, whereas humins are insoluble in water at all pH values [135-136]. Despite the arbitrary nature of this separation method, it is mostly used to achieve a segregation of humic substances according to their chemical properties, which chiefly depend on surface functional groups and macromolecular structure.

The major constituents of NOM are C, O, H, and N; other elements are P and S. The elemental compositions are more distinguishable according to fractionation (FA vs HA) than to sources and geographical areas. Fulvic acid have lower contents of C and N but higher contents of O and S than HA. Based on elemental compositions, the mole ratios of O/C, H/C, or N/C are calculated to indicate the degree of polarity/aromaticity and unsaturation of carbon chains. The O/C ratios are about 0.5 for HA and about 0.7 for FA. The H/C ratios, which have a larger variation, are about 1.0 for HA and about 1.4 for FA. In general, FA has greater aliphatic character, while HA is of higher aromaticity and hydrophobicity. Insoluble humin is believed to be similar to HA, with slightly less aromatic groups and acidity but higher contents of polysaccharide and alcoholic OH [132, 134, 135].

Suwannee River natural organic matter (SRNOM) is a well-known end member of NOM from an aquatic system and is a reference material of the International Humic Substances Society. The SRNOM was acquired from 36,890 L of filtered river water that was concentrated 40-fold on-site using reverse osmosis (RO) systems. After RO, the concentrated sample was desalted by cation exchange (CEX), freeze dried, and homogenized. The overall yield of dissolved organic carbon (DOC) was 84.2%. The final NOM sample supplied to the IHSS, which is designated 2R101N, contains only 3.89% inorganic ash, which reasonably allows most chemical analyses [127, 138, 139].

The work highlighted in this section adds to our understanding of the chemical interactions between NOM and U which affect U adsorption, precipitation, and dissolution in organic-rich environments at environmentally relevant pH. While the presence of NOM in sandstone formations is well established, little is known about the effects of uranium interaction with NOM on organic group functional chemistry [5, 6, 23, 140]. Improving the current understanding on the organic functional group chemistry found in sandstone formations and mineralized U deposits will help to identify the binding mechanisms influencing the interactions of U and NOM. These interactions may ultimately affect the reactive transport of U in soils, surface water and groundwater, and uptake in plants. Although several methods have been used to understand the mineralogy of sandstone formations and U mineralized deposits, the analysis of particulate NOM is challenging [126]. The application of X-ray photoelectron spectroscopy (XPS) to observe changes of organic functional groups in U mineralized deposits as a result of pH changes has not been widely used [141, 142].

The effect of pH on the precipitation of U in the presence of NOM needs to be further investigated given the effect it has on U and NOM chemistry. One approach has been modelling the reactivity of U with specific functional groups (e.g., carboxyl or phenolic) for fulvic acid (FA) and humic acid (HA) [6, 24, 129]. Few experimental studies have been conducted to understand the influence of pH on the precipitation of U in the presence of NOM [137, 143-146]. Laboratory experiments conducted in a previous study showed that the complexation of uranyl with carboxyl groups of NOM is possible by forming uranyl–carboxyl compounds [21]. Despite previous efforts, the role of pH on U and NOM reactions which ultimately influences U solubility in the environment needs to be better understood.

While there are several studies of the interactions between DOM and U, there is limited information about the reactions of U-DOM that control the complexation chemistry and its effect on U solubility in the environment [143, 147-149]. Dissolved organic matter contains metal-binding functional groups (such as carboxylates, phenols, amines, thiols) with binding affinities and ligand densities spanning many orders of magnitude [150]. Binding of U by DOM remains poorly understood at the molecular scale under environmentally relevant conditions [150]. Understanding the reaction mechanisms between U and DOM continue to be challenging at the molecular scale by the intrinsic complexity of DOM and U chemistry, the lack of binding constants, and analytical limitations [150]. Improving the current understanding of the influence of pH on U-DOM complexes will help to identify how the solubility of U changes in natural waters [87, 92, 148-149]. Although there is evidence that the hydrophobicity of DOM influences U mobility [151], the behavior of different functional groups as a function of pH and their reaction with U is still not well understood.

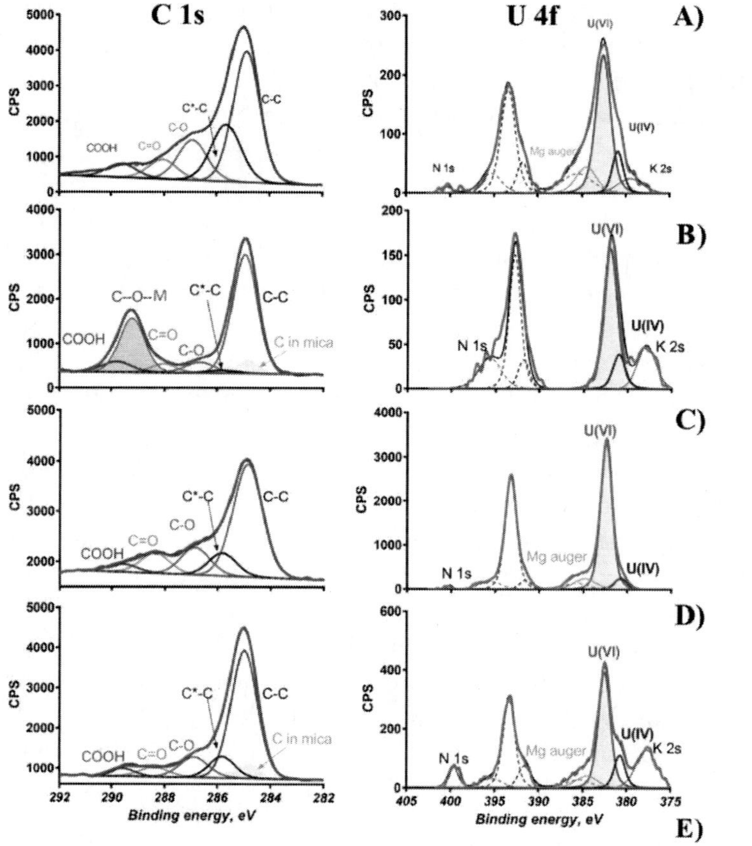

E)

Sample	pH	Carbonyl (C=O)	Phenol (COH)	Carboxylic (COOH)	C-O-M	U (IV)	U (VI)
Unreacted Mine Wastes	as is	8.6%	18.3%	5.8%	0.0%	19.4%	80.6%
	13	3.8%	5.4%	5.4%	23.5%	6.6%	93.4%
Supernatant	7	10.7%	14.4%	4.2%	0.0%	22.1%	77.9%
	2	5.3%	11.0%	3.7%	0.0%	31.0%	69.0%
	13	10.3%	14.6%	5.3%	0.0%	5.4%	94.6%
Reacted Solids	7	5.7%	12.7%	3.6%	0.0%	15.0%	85.0%
	2	2.2%	5.9%	5.2%	0.0%	4.8%	95.2%

Figure 7. Fitting of high resolution XPS Carbon (C 1s) and Uranium (U 4f) spectra of A) unreacted Jackpile Mine solids. Fitting of high resolution XPS Carbon (C 1s) and Uranium (U 4f) spectra of supernatant from batch extraction reactors at: B) pH 13; C) pH 7; and D) pH 2; and E) Percent composition of C 1s and U 4f spectra. Adapted with permission from Ref. [152]. Copyright 2019 American Chemical Society.

The integration of experiments using natural samples from mineralized deposits and laboratory-controlled experiments with advanced analytical techniques could provide new information about the effect of pH on interfacial reactions such as complexation, adsorption and precipitation between NOM, and U.

This research sought to advance the understanding of the interfacial chemical reactions between U and NOM in organic-rich environments which are currently not well understood. Part of this work contributes to the state of the current knowledge by identifying the organic functional group chemistry and U species (i.e., U(IV) and U(VI)) in mineralized deposits. Laboratory experiments aim to studying the effects of pH and NOM on the adsorption and precipitation, as well as the changes on the organic functional group chemistry of DOM due to these reactions.

Within these vignettes, we summarize results obtained by using natural samples from an organic-rich mineralized U deposit and laboratory-controlled conditions to investigate the chemical reactions affecting the speciation of U(IV) and U(VI), and the organic functional group chemistry in organic-rich environments using aqueous analyses, advanced spectroscopy, microscopy, and ultrahigh-resolution mass spectrometry. Specifically, we: A) discuss how changes in pH affect the binding of carboxylic functional groups in NOM and the speciation of U(IV) and U(VI) in mineralized deposits, B) investigate the pH-dependent precipitation of U(VI) and NOM, and C) identify changes in DOM chemistry due to the reaction of NOM and U at various pHs.

3.2. Vignette C: Organic Functional Group Chemistry in Mineralized Deposits Containing Uranium

In the first research vignette of this section, we explore the organic functional group chemistry of natural organic matter interacting with U and V in mine deposits. We investigated the functional group chemistry of natural organic matter (NOM) associated with both U(IV) and U(VI) in solids from mineralized deposits exposed to oxidizing conditions from the Jackpile Mine, Laguna Pueblo, NM. The uranium (U) content in unreacted samples was 0.44–2.6% by weight determined by X-ray fluorescence. In spite of prolonged exposure to ambient oxidizing conditions, ≈49% of U(IV) and ≈51% of U(VI) were identified on U LIII edge extended X-ray absorption fine structure spectra. Loss on ignition and thermogravimetric analyses identified from 13% to 44% of NOM in the samples. Carbonyl, phenolic, and carboxylic functional groups in the unreacted samples were identified by fitting of high-resolution X-ray photoelectron spectroscopy (XPS) C 1s and O 1s spectra. Peaks corresponding to phenolic and carbonyl functional groups had intensities higher than those corresponding to carboxylic groups in samples from the supernatant from batch extractions conducted at pH 13, 7, and 2. U(IV) and U(VI) species were detected in the supernatant after batch extractions conducted under oxidizing conditions by fitting of high-resolution XPS U 4f spectra. The outcomes from this study highlight the importance of the influence of pH on the organic functional group chemistry and U speciation in mineralized deposits. XPS results are summarized in Figure 7.

The main findings of this work indicated that phenolic and carbonyl functional groups are more abundant than carboxyl functional groups in the NOM from the Jackpile Mine mineralized deposit samples and show detectable changes on the XPS C 1s spectra after batch extraction at pH 13, pH 7, and pH 2. Contrary to what is commonly found in NOM [153, 154], carboxylic functional groups were the least abundant. Limited changes in the XPS C 1s peaks characteristic of the carboxylic groups were observed as a function of the pH conditions tested in this study. A unique finding of our study is that U(IV) was identified in unreacted solids, exposed to surface oxidizing conditions for several decades, as well as in reacted solids and in the supernatant collected from the batch reaction experiments conducted in oxidizing conditions. In addition to detecting U(IV) in supernatants from batch experiments at pH 13, 7, and 2; we observed noticeable differences in photoelectron peak intensities corresponding to phenolic- and carbonyl-peaks, in comparison to those for carboxylic-functional groups. The organic functional groups in these mineralized deposit samples from the Jackpile Mine could affect the redox, complexation, and precipitation chemistry of U. These chemical reactions could be relevant to the mobilization of the dissolved, colloidal, and particulate forms of U(IV) and U(VI) as shown in other studies [93, 155]. For instance, a recent investigation found high concentration of U(VI) in organic-rich layers as monodentate complexes bound to carboxyl and phosphoryl groups of humic substances above a wetland redox boundary, and the complexation of U(IV) by NOM in reducing conditions [156]. Understanding changes in organic functional chemistry at pH 7 and pH 2 is environmentally relevant due to their influence on chemical reactions that could impact U mobility and transport from the Jackpile Mine to surface waters and plants [5, 140]. For example, circumneutral pH conditions are characteristic of surface waters adjacent to the Jackpile Mine, while lower pH is commonly observed in sites affected by acid mine drainage [5, 157, 158].

Future research is necessary to identify specific mechanisms through which functional groups from the NOM found in our study control U reactivity in sites with complex mineralogy such as the Jackpile Mine, particularly the influence of NOM on U(IV) stabilization in oxic conditions. This information is essential for the development of risk assessment and remediation strategies in this site and other organic-rich uranium mineral deposits. These findings identify the need for further investigations related to the influence of organic functional groups on heavy metal transport and reactivity.

3.3. Vignette D: From Adsorption to Precipitation of U(VI) in the Presence of Natural Organic Matter

Natural organic matter can facilitate the precipitation of crystalline U-bearing solids under acidic pH conditions relevant to natural and anthropogenic environments. We investigated the interfacial reactions of U(VI) in the presence of natural organic matter (NOM) at pH 2, pH 4, and pH 7 in batch experiments with analyses done by chemical methods, electron microscopy, and advanced spectroscopy. Batch experiments reacted 200 mg L^{-1} of Suwannee River NOM with 100 μM U(VI) in a 0.01M KCl solution. Soluble U concentrations at pH 2 and pH 4 after 0.5 h and 24h of reaction (p-value <0.05) decreased after reaction with NOM for 0.5 h, followed by an increase to near the initial concentration after 24 h. Soluble U at pH 7 at 0.5 h and 24 h remained close to the initial concentration of 100 μM, indicating that NOM facilitates the solubility of U at circumneutral pH. Results of these batch experiments are summarized in Figure 8.

Bulk analyses conducted by EXAFS indicated that U(VI) is mainly adsorbed to particulate organic matter (POM) at pH 4. Precipitates of U(VI)- and K-bearing crystalline solids at pH 4 were also characterized by transmission electron microscopy (TEM). This investigation demonstrated that adsorption and precipitation of U(VI) in the presence of NOM is relevant at pH 2 and pH 4, while the aqueous complexation of U by dissolved organic matter (DOM) [160] at pH 7 prevents its adsorption and precipitation. The reactions of U(VI) with NOM are relevant to mineralized deposits, radwaste repositories, wetlands, and other environmental systems.

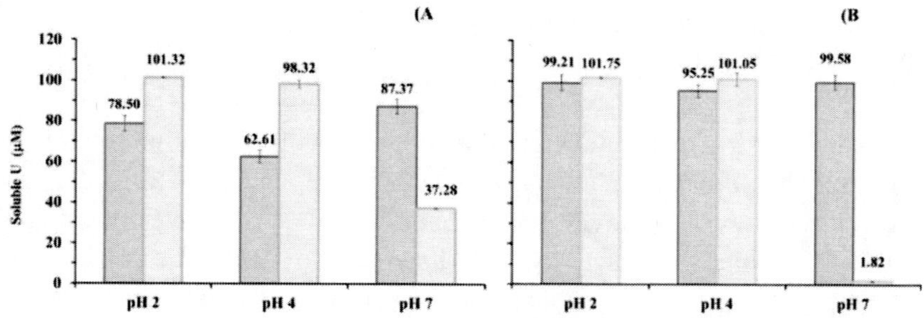

C)

Sampling Time	Soluble U (µM)					
	pH 2		pH 4		pH 7	
	U-KCl-NOM	U-KCl	U-KCl-NOM	U-KCl	U-KCl-NOM	U-KCl
After 0.5 hr.	78.5±10.8	101.3±5.3	62.6±19.3	98.3±0.5	87.4±3.3	37.3±13.1
After 24 hr.	99.2±3.7	101.7±1.8	95.3±3.1	101.1±1.8	99.6±3.7	1.8±0.7

Figure 8. Soluble U concentration in batch experiments containing NOM, U and KCl (left, darker bars), and U control experiment (right, lighter bars) at (A) 0.5, (B) 24 h, and (C) soluble U concentration summary results. Initial concentrations used are 200 mg L^{-1} NOM and 100 µM-$UO_2(NO_3)_2$ in 0.01 M KCl. Adapted with permission from Ref. [159]. Copyright 2021 American Chemical Society.

Results from this work reveal information about interfacial reactions of U and NOM in organic-rich environments using a well-characterized organic material, Suwanee River NOM. The influence of pH on U, POM, and DOM reactions can affect the solubility and accumulation of U in natural environments. Adsorption and precipitation were studied at a reaction time of 0.5 h and 24 h, which is applicable for short-term scenarios in natural and engineered systems. For example, the retention time in engineered systems for ion exchange and precipitation of U usually ranges from minutes to hours [161]. However, these time scales are much shorter than those occurring in natural environments such as an aquifer or involving organic detritus in the bottom of a wetland [131]. These reactions may occur in remediation processes such as treatment of acid mine drainage solutions (pH ranging between 2 and 4) and in natural waters (circumneutral pH). This work highlights the importance of metal and NOM interactions as these affect surface complexation, precipitation, and solubilization reactions that should be considered in reactive transport models for risk assessment and remediation purposes.

3.4. Vignette E: Effect of Oxidizing Conditions on Submicron U Phases in Mine Wastes

The final vignette of this section is a study of the mobility of U from submicron U(IV) and U(VI) mineral phases in the Jackpile-Paguate mine wastes under oxidizing conditions. We investigated the effect of bicarbonate and oxidizing agents on uranium (U) reactivity and subsequent dissolution of U(IV) and U(VI) mineral phases in the mineralized deposits from Jackpile mine, Laguna Pueblo, New Mexico, by integrating laboratory experiments with spectroscopy, microscopy, and diffraction techniques. Uranium concentration in solid samples from mineralized deposits obtained for this study exceeded 7000 mg kg^{-1}, as determined by X-ray fluorescence (XRF). Results from X-ray photoelectron spectroscopy (XPS) suggest the co-existence of U(VI) and U(IV) at a ratio of 19:1 at the near surface region of unreacted solid samples. Analyses made using X-ray diffraction (XRD) and electron microprobe detected the presence of coffinite ($USiO_4$) and uranium-phosphorous-potassium (U-P-K) mineral phases. Imaging, mapping, and spectroscopy results from scanning transmission electron microscopy (STEM) indicate that the U-P-K phases were encapsulated by carbon. Despite exposing the solid samples to strong oxidizing conditions, the highest aqueous U concentrations were measured from samples reacted with 100% air saturated 10 mM $NaHCO_3$ solution, at pH 7.5. Analyses using X-ray absorption spectroscopy (XAS) indicate that all the U(IV) in these solid samples were oxidized to U(VI) after reaction with dissolved oxygen and hypochlorite (OCl^-) in the presence of bicarbonate (HCO_3^-). The reaction between these organic rich deposits, and 100% air saturated bicarbonate solution (containing dissolved oxygen), can result in considerable mobilization of U in water, which has relevance to the U concentrations observed at the Rio Paguate across the Jackpile mine. Results from this investigation provide insights on the reactivity of carbon encapsulated U-phases under mild and strong oxidizing conditions that have important implication in U recovery, remediation, and risk exposure assessment of sites.

Based on results from XPS analysis, submicron U(VI) and U(IV) phases were found in solid samples from the Jackpile Mine at a 19:1 ratio.

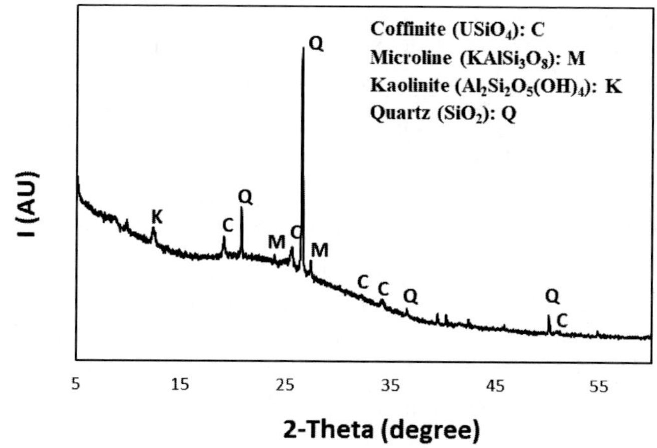

Figure 9. X-Ray Diffraction patterns of unreacted Laguna ore indicating presence of Coffinite (C, U(IV) phase), Microcline (M), Kaolinite (K), and Quartz (Q). These phases were found to have normalized weight percentages of 1.4, 2.6, 0.4, and 49%, respectively; the balance is 47% amorphous and other phases. Adapted from Ref. [162] with permission from Elsevier, 2019.

The results from this study provide relevant information about how the reactivity of submicron U-bearing mine wastes in the Jackpile-Paguate Mine can affect U mobility. For instance, elevated U concentrations (35.3 to 772 mg L^{-1}) have been measured at alkalinity and pH conditions (average pH 7.5, 70 – 372 mg L^{-1} $CaCO_3$) in the Rio Paguate just across the Jackpile-Paguate Mine [6]. The pH and alkalinity conditions selected for these experiments are relevant to the Rio Paguate. This information provides insights about the reactivity of U-micro particles in abandoned mine wastes into neighboring water resources to assess risks for human exposure.

Additionally, microscopy (Figure 10) and XRD (Figure 9) analyses on these mine wastes identified that these U-minerals occur predominantly as coffinites and U-P-K phases as the principal U solid phases that were encapsulated by organic carbon [6, 124, 131]. The association of U with different elements was indicative of the complex mineralogy of these solids. Results from mine waste leaching with $NaHCO_3$ and strong oxidants suggest that dissolution of U-phases and subsequent sorption and/or precipitation are key processes affecting U mobility. The highest U concentrations were mobilized on leaching with 10 mM $NaHCO_3$ solution containing ambient DO concentrations. These results partially explain the presence of elevated U concentrations in the Rio Paguate (35.3 to 772 mg L^{-1}) under relevant alkalinity and pH conditions (average pH 7.5, 70 – 372 mg L^{-1} $CaCO_3$) as reported by others [6]. This information can be useful to provide an improved understanding on the reactivity and mobility of U from submicron U minerals in abandoned mine wastes into neighboring water resources and to assess risks for human exposure. Knowledge from this study also provides ideas to improve mine waste management strategies through targetted U remediation and resource recovery initiatives. However, several gaps remain including 1) the role of organic matter in the reactivity and mobility of U from these submicron U minerals and 2) the reactvity and solubility of different U minerals under relevant field conditions. More research is necessary to overcome these research gaps to better understand the mechanisms that affect the mobility of U from organic rich mine wastes.

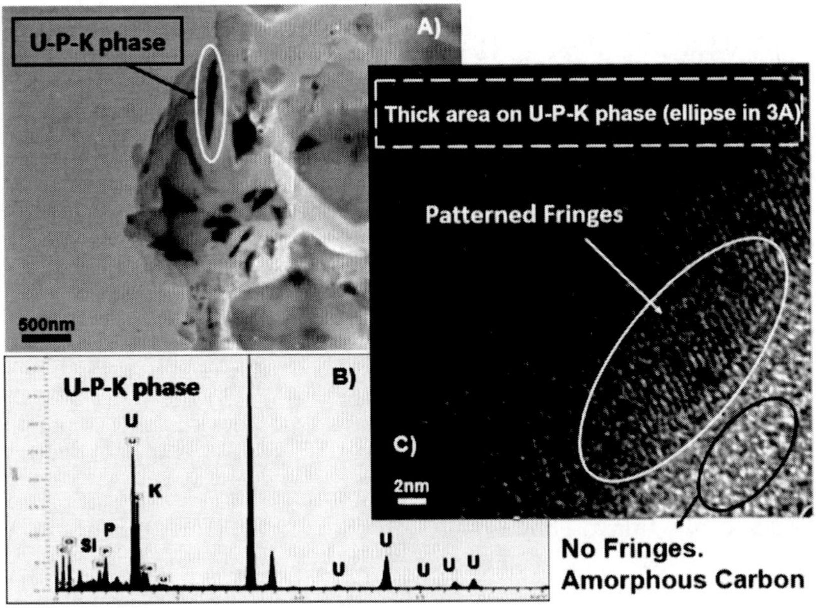

Figure 10. Scanning Transmission Electron Microscopy (STEM) imaging of U phases within unreacted ore. A) STEM image of U-P-K phases in black. B) Energy Dispersive Spectrum (EDS) on grain 1 (white ellipse in image 10A), confirming the presence of U-P-K phase. C) High resolution- transmission electron microscopy analysis on grain 1 (U-P-K phase) showing patterned fringes (in while ellipse in panel 10B) suggesting the phase to be crystalline. The lack of patterned fringes around the U-P-K phase (in black ellipse) suggests the presence of an amorphous phase possible organic matter in the ore sample. Adapted from Ref. [162] with permission from Elsevier, 2019.

These studies help to explain the role of thermodynamics, crystallinity, stoichiometry, and solution chemistry on the reactive transport of U and V from such mine wastes that pose as a threat to neighboring surface and ground water quality.

4. Mobility of Uranium and Arsenic in Mine Wastes

The goal of work in this section was to investigate the mechanisms of immobilization of uranium (U) and arsenic (As) in waters nearby to Native American Communities for remediation applications. We sought to identify the physical or chemical processes affecting the complexation, sorption, and precipitation of U and As with calcium (Ca), carbonates $(CO_3)^{2-}$, and phosphate $(PO_4)^{3-}$. The limited current knowledge about the mechanisms affecting the reaction of metal mixtures such as U, As, and other co-occurring elements in natural waters with minerals such as limestone and hydroxyapatite, and fluorapatite [163-165] for remediation applications is first reviewed before detailing our work. The research detailed in these vignettes summarizes our work to:

- Vignette F) Determine the solubility product (log K_{sp}) and thermodynamic variables ($\Delta G°$, $\Delta H°$, $\Delta S°$) for uranyl arsenate solids with monovalent cations sodium (Na) and potassium (K). Specifically, the behavior of potassium uranyl arsenate solid $[K(UO_2)(AsO_4)](H_2O)_3]$ and sodium uranyl arsenate $[Na(UO_2)(AsO_4)](H_2O)_3]$ are compared.
- Vignette G) Determine the kinetic rate law for dissolution of uranyl arsenate solids at acidic pH. $NaUAs_{(s)}$ and $KUAs_{(s)}$ were again compared and expected to have similar dissolution rates.

4.1. State of Knowledge about U and As Co-Occurrence in Mining Legacy Sites and the Kinetics and Thermodynamics of Uranyl Arsenate Interactions

Both uranium (U) and arsenic (As) naturally occur and are commonly associated in uranium deposits [34]. The most relevant oxidation states for U in natural environments are U(VI) and U(IV) [166], though U can exist in nature in oxidation states ranging from +4 to +6 [167] and it occurs as oxides, peroxides, oxide-hydroxy-hydrates, precipitates, and complexes [167]. Uranium speciation and mobility depend on the presence of dissolved inorganic carbon and organic matter, the solution pH, redox potential, hydrolysis, dissolution, complexation, and sorption [5]. Uranium (IV) exists in reduced subsurface environments and is less mobile than U(VI). Uranium (VI) is the most prevalent oxidation form in the environment, and occurs in oxic conditions across the pH range, forming more soluble and toxic compounds compared to U(IV) [166]. The U(VI) cation in the environment is invariably present as a $(UO_2)^{2+}$ uranyl ion.

The predominant oxidation states of As in natural waters are As(III) and As(V) [168], but As can also occur in nature in oxidation states ranging from -3 to +5. Arsenic speciation also depends on pH and redox potential of the environment [169]. For instance, As(V) predominates in oxidizing conditions and pH values lower than 6.9 as H_2AsO_4, in higher pH values as $HAsO_4^{2-}$, in extremely acidic conditions as H_3AsO_4, and extreme alkaline conditions as AsO_4^{3-}. As(III), on the other hand, will predominate under reducing conditions and pH values lower than 9.2 mainly as H_3AsO_3 [168]. Both As(V) and As(III) are toxic; however, As(III) is 20 to 50 times more toxic than As(V) [169].

The co-occurrence of U and As caused by natural and anthropogenic processes is a concern to communities living in the proximity of contaminated sites [34-40]. Minerals containing U and As occur in various locations worldwide [40-42] and communities that surround the mine sites have been impacted by the release of U and As, which is of major concern for public health as both are toxic and carcinogenic, and U is radioactive [34, 43-44]. As previously discussed, exposure to U can lead to chemical and radiological toxicity [166]. Its main chemical effect is associated with kidney disease while its primary radiological effect is associated with cancer [170]. Research evaluating the effect of uranium on Navajo communities have related U exposure to an increased probability of developing one or more chronic diseases such as hypertension, kidney disease, and diabetes [51]. On the other hand, exposure to As can lead to its accumulation in skin, hair, and nails resulting in clinical conditions such as hyperpigmentation or keratosis. It also increases the risk of skin, lung, and internal organ cancer. In addition, other conditions such as cardiovascular diseases, neuropathy, low verbal IQ, loss of long term memory, fetal loss and premature delivery, and decreased birth weights of infants have been associated with exposure to As [171]. Consequently, reclamation and environmental restoration of abandoned U mine sites are now being addressed under the Superfund law by the USEPA, state agencies, and tribes [20].

Considering areas not influenced directly by mining operations, U and As can co-occur as contaminants and impact human health as well. For example, the spatial relationship between U and As in Northern Plains Native lands were evaluated and U and As were found elevated in samples taken from three regions in North and South Dakota [38]. Additionally, a spatial correlation was seen between ground water wells containing U and As which are associated with flow along a geologic contact; however, the underlying mechanisms enabling the U and As correlation warranted further study [38]. Therefore, it is crucial to develop a greater understanding of U and As and their behavior in natural waters.

Uranyl arsenate solids ($UAs_{(s)}$) are one of the most diverse chemical groups of U minerals but have been less characterized than the analogous phosphate phases [41, 172]. Arsenate and phosphate-bearing anionic units are generally isostructural, have similar size and charge, precipitate as uranyl minerals at elevated concentrations, and form ternary complexes on surfaces at low concentrations [34, 173-176]. Uranyl phosphates have generally low solubilities ranging from log K_{sp} of -13.17 for uranyl hydrogen phosphate to -49.36 and -50.39 for uranyl orthophosphate [176-177]. The solubility curves and solubility products (log K_{sp}) of $UAs_{(s)}$ like $(UO_2)_3(AsO_4)_2 12H_2O$, $NaAsUO_6 3H_2O$, and $KAsUO_6 3H_2O$ have been measured [176, 178]. However, these previous measurements were done only within heterogenous water-salt systems or saturated conditions [176, 177], and literature suggests that solubility measurements should be confirmed under both supersaturation and undersaturation [176, 177-178]. The changes in U and As soluble concentration as a function of pH could also be affected by the formation of secondary phases. Secondary phases that have been reported in the literature, include trögerite $(UO_2)_3(AsO_4)_2 12H_2O_{(s)}$, [176, 181] and $(UO_2)(H_2AsO_4)_2(H_2O)_{(s)}$ [182, 183]. These previous studies indicate that the thermodynamics and solubility constants of $UAs_{(s)}$ need more documentation [182, 184].

Understanding the kinetics of the dissolution of UAs(s) is also essential to gain fundamental knowledge of U and As mobilization in environmentally relevant conditions. This knowledge is necessary for the use of reactive transport models as a tool to interpret and predict the fate of these toxic elements in the environment and to develop risk assessment and remediation strategies.

Active and passive treatments are remediation strategies for treating acid mine drainage (AMD) and waste of abandoned mine sites. Active treatments involve the application of alkaline materials to remediate acidic mine waters and precipitate metals, while passive treatments use the application of natural and constructed wetlands [185]. Limestone is a sedimentary rock mainly composed of calcium carbonate ($CaCO_3$) commonly applied in both passive and active treatments for the neutralization of acid mine drainage [186]. Passive treatments that depend exclusively on the application of limestone are commonly used in post-closure/low acidic mine sites and are relatively inexpensive and easy to maintain [186].

Several studies have evaluated the use of limestone to remove heavy metals from contaminated water [187-190]. Limestone (95.6% $CaCO_3$) showed removal efficiencies higher than 82% for cadmium (Cd), copper (Cu), chromium (Cr), lead (Pb), nickel (Ni), and zinc (Zn) [187]. Limestone columns (95% $CaCO_3$) removed 99% of As [188]. Limestone (77 and 99% $CaCO_3$) exhibited removal efficiencies higher than 88% for Cd, Cu, and Zn [189]. Finally, limestone has also shown removal efficiencies for Pb as high as 90% [190]. However, very few studies have assessed the application of limestone to treat mine waste containing U in co-occurrence with As.

Research has indicated that there are various processes involved in the sorption mechanisms of heavy metals in limestone [189-191]. Earlier findings suggest that at a low pH, the rapid dissolution of calcium carbonate is the first mechanism for heavy metal sorption forming carbonate complexes with limestone [189]; for example, Godelitsas et al., [190] showed formation of $PbCO_3$. Labastida et al., [191] indicated that metal removal also occurs due to precipitation and co-precipitation of metals. However, other results have indicated that the presence of some heavy metals such as Cd inhibited the dissolution of carbonate [$(CO_3)^{2-}$] to some extent, suggesting that there are other mechanisms involved in the removal of metals by limestone [189]. Iron hydroxides have also decreased the $(CO_3)^{2-}$ dissolution from limestone grains [192]. Overall, these studies indicate that the mechanism involved in the sorption of metals in limestone are complex and metal-specific. The presence of co-occurring U and As may likely have a unique sorption mechanism in a limestone matrix.

Precipitation of U and As in water will also depend on the working pH. Environmentally relevant conditions limit remediation approaches based on adsorption and precipitation with natural minerals. The formation of ternary uranyl aqueous complexes such as $Ca_2UO_2(CO_3)_3^0{}_{(aq)}$ increase U mobility [42, 75]. In a previous study it was observed that $UAs_{(s)}$ are stable in aqueous solution at pH 3.0 but at pH 7.0 dissolution is enhanced by formation of ternary calcium (Ca) uranyl carbonate complexes [193]. The availability of environmentally relevant ions in solution, like bicarbonate and $(PO_4)^{3-}$, can release $(AsO_4)^{3-}$ through competitive ion displacement [194]. Research has also found that $(UO_2)^{2+}$ and $(AsO_4)^{3-}$ can decrease in concentration when amendments with $(PO_4)^{3-}$, and both limestone and apatite have been used in batch experiments at pH 3.0 and 7.0.

The surface charge of any abundant mineral like calcite or apatite has to be considered when U and As are removed from aqueous solutions. Zeta potential analyses indicated the limestone used in this study is negatively charged from pH 2.0 to 9.0 at 25°C. Considering previous research, if the limestone is made up of quartz, albite, and calcite, it might have a negative charge that as well will influence U and As immobilization from water through adsorption and surface precipitation [195-197].

4.2. Vignette F: Solubility and Thermodynamics of Monovalent Uranyl Arsenates

Within this first research vignette for this section, we summarize our investigation into the aqueous solubility and thermodynamic properties of two meta-autunite group uranyl arsenate solids (UAs). The aqueous solubilities and calorimetric measurements for monovalent uranyl arsenates provided thermodynamic data necessary for modeling mobilities of uranium and arsenic in the environment. The measured solubility products (log K_{sp}) obtained in dissolution and precipitation experiments at equilibrium pH 2 and 3 for NaUAs and KUAs ranged from −23.50 to −22.96 and −23.87 to −23.38, respectively. The secondary phases $(UO_2)(H_2AsO4)_2(H_2O)(s)$ and trögerite, $(UO_2)_3(AsO_4)_2 \cdot 12H_2O(s)$, were identified by powder X-ray diffraction in the reacted solids of KUA precipitation experiments (pH 2) and NaUAs dissolution and precipitation experiments (pH 3), respectively. The identification of these secondary phases in reacted solids suggest that H_3O^+ co-occurring with Na or K in the interlayer region can influence the solubilities of uranyl arsenate solids. The standard-state enthalpy of formation from the elements ($\Delta H_{f\text{-}el}$) of NaUAs is −3025 ± 22 kJ mol^{-1} and for KUAs is −3000 ± 28 kJ mol^{-1} derived from measurements by drop solution calorimetry, consistent with values reported in other studies for uranyl phosphate solids.

Uranyl arsenate solids (NaUAs and KUAs) were synthesized by the diffusion method as described by Dzik [34]. This method is known for providing high quality, purity, and yield of the material required for calorimetric measurements. Chemical analyses, p-XRD, and microscopy indicated that single phases of NaUAs and KUAs were obtained (see Figure 11). Single phase crystals of uranyl arsenates of the meta-autunite group with mono and divalent cations have also been identified in previous studies after synthesis using the same method [34, 198]. Chemical analyses of both compounds NaUAs and KUAs indicate that H_3O^+ is present with K^+ or Na^+ in the interlayer region of the structures. The hydronium-bearing uranyl arsenate analog of uranospinite, $(H_3O)_2[(UO_2)_2(AsO_4)_2](H_2O)_8$, has been documented [199]. Chernikovite is a well-characterized H_3O^+-bearing autunite-structure compound with composition $(H_3O)_2[(UO_2)_2(PO_4)_2](H_2O)_n$, and a hydronium-$K^+$ solid-solution series has been noted in this uranyl phosphate system [200, 201]. Hydronium substitution for various monovalent cations in autunite-type uranyl compounds has also been documented [202-204]. Furthermore, a reported single-crystal structure determination for KUAs assigned the formula $K_{0.91}(H_3O)_{0.09}[(UO_2)(AsO_4)](H_2O)_3$ [198], similar to that seen in our study.

Measurements from dissolution and precipitation experiments, in combination with thermochemical modelling with PFLOTRAN, allowed us to determine solubility product constants for the uranyl arsenates at various pH values. The solubility product constants obtained at pH 2 and 3 for KUAs and NaUAs are within one order of magnitude, which is within expected experimental error for soluble measured concentrations. The substitution of cations are known to impact the number of H_2O molecules in the interlayer of UAs [175]. Uranium and As form anionic two-dimensional layers of the type $[P(As)UO_6]^{n-}{}_{2\infty}$ known by an increased stability [175]. For instance, the role of hydronium in the UAs structure needs to be further understood.

The occurrence of hydronium $(H_3O)^+$ in the interlayer region of compounds containing uranyl phosphate and arsenate sheets with the autunite topology is lacking an understanding of the details of the substitution of hydronium for other monovalent cations in the interlayers of autunite-structure compounds. In the case of autunite-structure compounds, the importance of hydronium is well established. The structure of the hydronium end-member compound $(H_3O)[(UO_2)(PO_4)](H_2O)_3$ was reported by Morosin and is a hydrogen ion solid electrolyte [206]. The structure of $(H_3O)[(UO_2)(AsO_4)](H_2O)_3$ was determined more recently, including the location of the H atoms corresponding to the hydronium ions [207]. Earlier determinations of the structures of autunite compounds containing monovalent cations also noted the presence of hydronium [198]. However, we are missing more information about the substitution of hydronium for monovalent cations in the interlayers of autunite-structure compounds, as hydronium is rare in mineral structures, and it is challenging to identify it.

The presence of hydronium is difficult to directly demonstrate by spectroscopy, and in the case of chemically well-characterized materials, its presence can be demonstrated by the electroneutrality principle. Locally, in the interlayer region, hydronium could substitute for a H_2O group or a monovalent cation, or the H atom may be largely mobile, which is consistent with the compound $(H_3O)[(UO_2)(PO_4)](H_2O)_3$ that exhibits rapid H^+ ion conductivity [208]. In the current case, the hydronium ion content was determined by establishing the composition of the compounds by chemical analyses, with hydronium assumed for charge balance. The identification of trögerite in the residual solids may relate to the role of H_3O^+ which is present with Na in the interlayer region of the structure of NaUAs used in the dissolution experiments; trögerite was also identified in precipitation experiments with NaUAs at pH 3. Formation of trögerite in this study caused a decrease in the final soluble U and As concentrations in solution and with a resulting final solution pH in the range of 3. The formation of this secondary mineral is consistent with findings reported in other studies [178, 181].

The aqueous speciation of U, As, Na, and K influences the dissolution and precipitation reactions for the determination of the log K_{sp} values of NaUAs and KUAs solids. For example, the formation of uranyl arsenate complexes has been reported in the literature and cannot be ignored in the determination of the solubility constants. According to the literature, at pH 2 and 3, U(VI) will be present as UO_2^{2+}, while As(V) could be present as 50% to 75% of H_3AsO_4 and $H_2AsO_4^{1-}$, and Na could be a free ion Na^+ [209]. However, we considered other aqueous complexes in the dissolution reactions with the PFLOTRAN modeling such as $UO_2H_2AsO_4{}^+{}_{(aq)}$, $UO_2HAsO_{4(aq)}$, and $UO_2(H_2AsO_4)_2$ [183, 184]. The log K_{sp} was not sensitive to these aqueous uranyl complexes as they differed only between 2.06% and 2.43%.

The solubility and thermodynamic data obtained for NaUAs and KUAs in this study contribute information that can underpin predictions of U and As dissolutions using chemical equilibrium and reactive transport models. The thermodynamic variables obtained can be used in reactive transport models to interpret and predict the solubilities of U and As. The findings from this study can be useful for risk assessment and risk reduction studies where anthropogenic activities such as mining and nuclear weapons development have resulted in contamination.

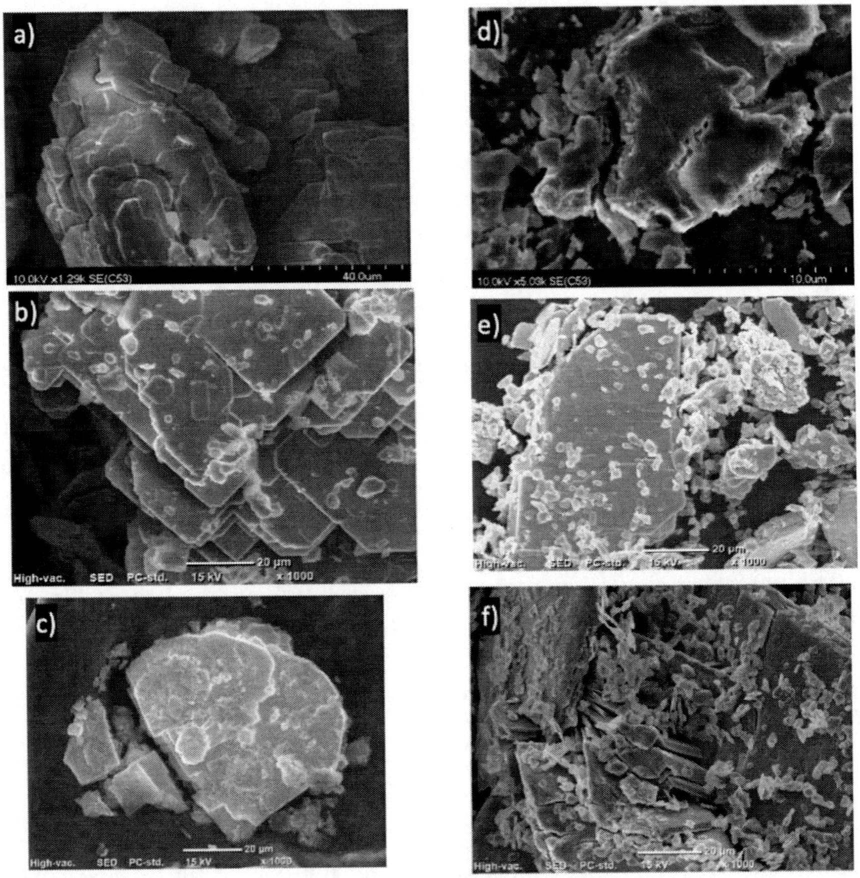

Figure 11. a) SEM image of unreacted NaUAs crystals. b) SEM image of reacted NaUAs from supersaturated experiments at pH 2. c) SEM image of reacted NaUAs from supersaturated experiments at pH 3. d) SEM image of unreacted KUAs crystals. e) SEM image of reacted KUAs from supersaturated experiments at pH 2. f) SEM image of reacted KUAs from supersaturated experiments at pH 3. Adapted with permission from Ref. [205]. Copyright 2023 American Chemical Society.

4.3 Vignette G: Insights into Kinetics and Chemical Equilibrium of Uranyl Arsenate Solids

Within the work described in this vignette, we integrated aqueous chemistry analyses with geochemical modeling to determine the kinetics of the dissolution of Na and K uranyl arsenate solids ($UAs_{(s)}$) at acidic pH. Improving our understanding of how $UAs_{(s)}$ dissolve is essential to predict transport of U and As, such as in acid mine drainage. Uranyl arsenates dissolve significantly faster than analogous uranyl phosphates. At pH 2, $Na_{0.48}H_{0.52}(UO_2)(AsO_4)(H_2O)_{2.5(s)}$ ($NaUAs_{(s)}$) and $K_{0.9}H_{0.1}(UO_2)(AsO_4)(H_2O)_{2.5(s)}$ ($KUAs_{(s)}$) both dissolve with a rate constant of 3.2×10^{-7} mol m^{-2} s^{-1}, which is faster than analogous uranyl phosphate solids. K- and Na- uranyl arsenates dissolve slower at pH 3 than pH 2. At pH 3, $NaUAs_{(s)}$ (6.3×10^{-8} mol m^{-2} s^{-1}) and $KUAs_{(s)}$ (2.0×10^{-8} mol m^{-2} s^{-1}) have smaller rate constants. Steady-state aqueous concentrations of U and As are similarly reached within the first several hours of reaction progress. This study provides dissolution rate constants for $UAs_{(s)}$, which may be integrated into reactive transport models for risk assessment and remediation of U and As contaminated waters.

The experiments in this study were conducted at acidic conditions, avoiding reactions with carbonate that would affect the formation of various secondary phases and aqueous complexes influencing the solubility of the minerals investigated here. The dissolution of $UAs_{(s)}$ at circumneutral pH could be enhanced by formation of strong aqueous uranyl-carbonate complexes [210]. Carbonate may lead to the dissolution of uranyl arsenate, uranyl phosphate, oxyhydroxide, and uranyl vanadate minerals as previously reported [14, 42, 113, 162, 211]. This study instead focused on the solubility of $UAs_{(s)}$ under acidic conditions, as found in acid mine drainage [212].

Our findings indicate that at pH 2.0 the presence of $NaUAs_{(s)}$ and $KUAs_{(s)}$ control the aqueous concentrations of U and As through their dissolution kinetics over the first hour. At pH 3.0, $KUAs_{(s)}$ rapidly dissolves and reaches chemical equilibrium. These results agree with previous studies which found that U and As in solution can be removed at acidic pH, with the precipitation of U-As-bearing minerals [193]. As a whole, our study provides insights about U and As mobilization and information for possible remediation approaches.

The results obtained in this study revealed that the concentrations of U and As are controlled by the reaction rates of Na- and K- $UAs_{(s)}$ at pH 2.0 (3.16×10^{-7} mol m^{-2} s^{-1}). Even though, $NaUAs_{(s)}$ dissolves slightly slower (7.94×10^{-8} mol m^{-2} s^{-1}) at pH 3 over the first hour of reaction but has trögerite as a secondary precipitate that impacts U and As concentrations in an aqueous

$NaUAs_{(s)}$ system. This difference indicates that reaction rates are pH dependent and the formation of secondary phases such as trögerite may decrease the reaction rate of $UAs_{(s)}$. For $KUAs_{(s)}$ at pH 3.0, U and As dissolve rapidly with the concentrations of U and As controlled by the chemical equilibrium for $KUAs_{(s)}$. The kinetic data obtained for $NaUAs_{(s)}$ and $KUAs_{(s)}$ contribute information that can underpin prediction of U and As concentrations at chemical equilibrium and/or kinetics in reactive transport models.

5. Further Reading, Attribution, and Acknowledgments

More information about the research vignettes detailed here can be found in the dissertations and papers as listed below:

Section 2.1 – https://digitalrepository.unm.edu/ce_etds/205/.
Avasarala, Sumant, "Physical and Chemical Interactions Affecting U and V Transport from Mine Wastes" Dissertation, University of New Mexico, Mar. 2, 2018.

Vignette A – https://pubs.acs.org/doi/abs/10.1021/acs.est.7b03823.
Selections from "Avasarala, S., Lichtner, P. C., Ali, A. M. S., González-Pinzón, R., Blake, J. M., Cerrato, J. M., Reactive Transport of U and V from Abandoned Uranium Mine Wastes. *Environ. Sci. Technol.* 2017, 51, 12385-12393." with permission of American Chemical Society.

Vignette B – https://www.mdpi.com/2075-163X/10/10/883.
Selections from "Avasarala, S., Brearley, A. J., Spilde, M., Peterson, E., Jiang, Y. B., Benavidez, A., Cerrato, J. M. Crystal Chemistry of Carnotite in Abandoned Mine Wastes. *Minerals* 2020, 10, 883." under Creative Commons BY 4.0 license.

Section 3.1 – https://digitalrepository.unm.edu/ce_etds/253/.
Velasco, Carmen A., "Chemical interfacial reactions affecting the speciation and solubility of uranium in organic-rich environments" Dissertation, University of New Mexico, Nov. 24, 2020.

Vignette C – https://pubs.acs.org/doi/abs/10.1021/acs.est.9b00407.
Velasco, C. A., Artyushkova, K., Ali, A. M. S., Osburn, C. L., Gonzalez-Estrella, J., Lezama-Pacheco, J. S., Cabaniss, S. E., Cerrato, J. M., Organic Functional Group Chemistry in Mineralized Deposits Containing U(IV) and U(VI) from the Jackpile Mine in New Mexico. *Environ. Sci. Technol.* 2019, 53, (10), 5758-5767.

Vignette D – https://pubs.acs.org/doi/abs/10.1021/acs.est.1c05429.
Selections from "Velasco, C. A., Brearley, A. J., Gonzalez-Estrella, J., Ali, A. M. S., Meza, M. I., Cabaniss, S. E., Thomson, B. M., Forbes, T. Z., Lezama Pacheco, J. S., Cerrato, J. M. From Adsorption to Precipitation of U(VI): What is the Role of pH and Natural Organic Matter? *Environ. Sci. Technol.* 2021, 55, 16246-16256." with permission from American Chemical Society.

Vignette E – https://doi.org/10.1016/j.chemgeo.2019.07.007.
Selections from "Avasarala, S., Torres, C., Ali, A. M. S., Thomson, B. M., Spilde, M. N., Peterson, E. J., Artyushkova, K., Dobrica, E., Lezama-Pacheco, J. S., Cerrato, J. M., Effect of bicarbonate and oxidizing conditions on U(IV) and U(VI) reactivity in mineralized deposits of New Mexico. *Chemical Geology* 2019, 524, 345-355." with permission from Elsevier.

Section 4.1 – https://digitalrepository.unm.edu/ce_etds/289/.
Meza, Maria I., "Thermodynamic and kinetic investigation of U(VI) and As(V) reactions: insights for risk assessment and remediation" Dissertation, University of New Mexico, Sept. 15, 2022.

Vignette F – https://doi.org/10.1021/acs.est.2c06648.
Selections from "Meza, I., Gonzalez-Estrella, J., Burns, P. C., Rodriguez, V., Velasco, C. A., Sigmon, G. E., Szymanowski, J. E. S., Forbes, T. Z., Applegate, L. M., Ali, A. M. S., Lichtner, P., Cerrato, J. M. Solubility and Thermodynamic Investigation of Meta-Autunite Group Uranyl Arsenate Solids with Monovalent Cations Na and K. *Environ. Sci. Technol.* 2023 57, 255-265." with permission from American Chemical Society.

Vignette G - https://doi.org/10.1016/j.chemgeo.2023.121642.
Selections from "Meza, I., Jemison, N., Gonzalez-Estrella, J., Burns, P. C., Rodriguez, V., Sigmon, G. E., Szymanowski, J. E. S., Ali, A. M. S., Gagnon, K., Cerrato, J. M., Lichtner, P. Kinetics of Na- and K- uranyl arsenate dissolution, *Chemical Geology*, 636, 2023, 121642." with permission from Elsevier.

The material included in these chapters is a result of the PhD dissertations of María I. Meza, Carmen Velasco, and Sumant Avasarala. This work was possible thanks to our various collaborators: Johanna M. Blake, Chris Torres, Abdul-Mehdi S. Ali, Michael N. Spilde, Adrian Brearley, Peter C. Lichtner, Juan Lezama, Ricardo Gonzalez-Pinzón, Kateryna Artyushkova, Angelica Benavidez, Ying-Bing Jiang, Stephen Cabaniss, Christopher L. Osburn, Jorge Gonzalez-Estrella, Eric Peterson, Noah Jemison, Peter Burns, Virginia Rodriguez, Ginger Sigmon, Jennifer Symanowski, Tori Forbes, Lindsey Applegate, Kaelin Gagnon, and Bruce Thomson. Funding for this research was provided by the National Science Foundation (awards NM EPSCoR #IIA-

1301346, CREST 1914490, and CAREER 1652619) and the National Institute of Environmental Health Sciences Superfund Research Program (Award 1 P42 ES025589). Any opinions, findings, and conclusions or recommendations expressed in this book chapter are those of the author(s) and do not necessarily reflect the views of the National Science Foundation or the National Institutes of Health.

Figures in this chapter have undergone graphical alterations/edits to conform to publication standards and are intended for use in this chapter as *illustrations*. The original referenced figures should be used for rigorous scientific examination/comparison.

References

[1] Riley, R. G., Zachara, J. "Chemical contaminants on DOE lands and selection of contaminant mixtures for subsurface science research"; Pacific Northwest Lab., Richland, WA (United States): 1992.

[2] Archer, V. E., Hazards to health in uranium mining and milling. *Journal of Occupational and Environmental Medicine* 1962, *4* (2), 55-60.

[3] Byrne, J., Hoffman, S. M., *Governing the Atom: The Politics of Risk.* North Atlantic Books: 1996; Vol. 7.

[4] Sovacool, B. K., The costs of failure: a preliminary assessment of major energy accidents, 1907–2007. *Energy Policy* 2008, *36* (5), 1802-1820.

[5] Bureau of Land Management, "Abandoned Mine Lands Portal," on-line, www.abandonedmines.gov, United States government 2016.

[6] Blake, J. M., De Vore, C. L., Avasarala, S., Ali, A. M., Roldan, C., Bowers, F., Spilde, M. N., Artyushkova, K., Kirk, M. F., Peterson, E., Rodriguez-Freire, L., Cerrato, J. M., Uranium mobility and accumulation along the Rio Paguate, Jackpile Mine in Laguna Pueblo, NM. *Environ. Sci-Proc. Imp.* 2017, *19*, (4), 605-621.

[7] Blake, J. M., Avasarala, S., Artyushkova, K., Ali, A. M. S., Brearley, A. J., Shuey, C., Robinson, W. P., Nez, C., Bill, S., Lewis, J., Elevated concentrations of U and co-occurring metals in abandoned mine wastes in a northeastern Arizona Native American community. *Environmental science & technology* 2015, *49* (14), 8506-8514.

[8] Chenoweth, W. "Vanadium mining in the Carrizo Mountains," 1942–1947. San Juan County, New Mexico and Apache County Arizona; New Mexico Bureau of Mines and Mineral Resources, 1991

[9] USEPA, "Radionuclides Rule: A Quick Reference Guide." Office of Water (4606). In EPA 816-F-01-003: 2001.

[10] USEPA, "Addressing uranium contamination on the Navajo Nation." U.S. Environmental Protection Agency, Pacific Southwest, Region 9: San Francisco, CA. 2015.

[11] USEPA, "Federal actions to address impacts of uranium contamination in the Navajo Nation: Five-year plan summary report." U.S. Environmental Protection Agency, Pacific Southwest, Region 9. San Francisco, CA 2013.

[12] Moore-Nall, A., The legacy of uranium development on or near Indian reservations and health implications rekindling public awareness. *Geosciences* 2015, *5*, (1), 15-29.

[13] USEPA, "Technical Report on Technologically Enhanced Naturally Occurring Radioactive Materials from Uranium Mining Vol. 1: Mining and Reclamation Background;" U.S. Environmental Protection Agency, Office of Radiation and Indoor Air, EPA 402-R-08-005:. Washington*, D.C.* 2008.

[14] U.S. Congress, "Uranium Mill Tailings Radiation Control Act of 1978." *Public Law 95-604*, Nov. 8, 1978.

[15] deLemos, J. L., Bostick, B. C., Quicksall, A. N., Landis, J. D., George, C. C., Slagowski, N. L., Rock, T., Brugge, D., Lewis, J., Durant, J. L. Rapid dissolution of soluble uranyl phases in arid, mine-impacted catchments near Church Rock, NM *Environ. Sci. Technol*. 2008, *42*, (11), 3951– 3957.

[16] Longsworth, S. A. "Geohydrology and water chemistry of abandoned uranium mines and radiochemistry of spoil-material leachate, Monument Valley and Cameron areas, Arizona and Utah." U.S. Geological Survey: Reston, VA, 1994.

[17] Nash, J. T., Uranium deposits in the Jackpile Sandstone, New Mexico. *Economic Geology* 1968, *63* (7), 737-750.

[18] USEPA, "Navajo Nation Cleaning Up Abandoned Uranium Mines," on-line, epa.gov/navajo-nation-uranium-cleanup, United States government 2024.

[19] USEPA, "Annual Progress Update: Federal Actions to Address Uranium Contamination on Navajo Nation." U.S. Environmental Protection Agency, Pacific Southwest, Region 9. San Francisco, CA Oct. 2023.

[20] ATSDR, "Public Health Assessment for Jackpile-Paguate Uranium Mine Laguna Pueblo, NM." U.S. Department of Health and Human Services, July 24, 2018.

[21] Zychowski, K. E., Kodali, V., Harmon, M., Tyler, C. R., Sanchez, B., Ordonez Suarez, Y., Herbert, G., Wheeler, A., Avasarala, S., Cerrato, J. M., Kunda, N. K., Muttil, P., Shuey, C., Brearley, A., Ali, A. M., Lin, Y., Shoeb, M., Erdely, A., Campen, M. J., Respirable Uranyl-Vanadate-Containing Particulate Matter Derived From a Legacy Uranium Mine Site Exhibits Potentiated Cardiopulmonary Toxicity. *Toxicological Sciences* 2018, *164*, (1), 101-114.

[22] Lewis, J., Hoover, J., MacKenzie, D., Mining and Environmental Health Disparities in Native American Communities. *Curr Environ Health Rep* 2017, *4*, (2), 130-141.

[23] Harmon, M. E., Lewis, J., Miller, C., Hoover, J., Ali, A. S., Shuey, C., Cajero, M., Lucas, S., Zychowski, K., Pacheco, B., Erdei, E., Ramone, S., Nez, T., Gonzales, M., Campen, M. J., Residential proximity to abandoned uranium mines and serum inflammatory potential in chronically exposed Navajo communities. *J Expo Sci Environ Epidemiol* 2017, *27*, (4), 365-371.

[24] Dawson, S., Navajo uranium workers and the effects of occupational illnesses: A case study. *Human Organization* 1992, *51* (4), 389-397.

[25] Hettiarachchi, E., Paul, S., Cadol, D., Frey, B., Rubasinghege, G., Mineralogy Controlled Dissolution of Uranium from Airborne Dust in Simulated Lung Fluids (SLFs) and Possible Health Implications. *Environ. Sci. Technol. Let.* 2018.

[26] Hopenhayn, C., Arsenic in drinking water: Impact on human health. *Elements* 2006, *2*, 103-107.

[27] D'amore, J., Al-Abed, S., Scheckel, K., Ryan, J., Methods for speciation of metals in soils. *Journal of environmental quality* 2005, *34* (5), 1707-1745.

[28] Byczkowski, J. Z., Kulkarni, A. P., Oxidative stress and pro-oxidant biological effects of vanadium. *Advances in Environmental Science and Technology -New York-* 1998, *31*, 235-264.

[29] Stohs, S., Bagchi, D., Oxidative mechanisms in the toxicity of metal ions. *Free radical biology and medicine* 1995, *18* (2), 321-336.

[30] Panichev, N., Mandiwana, K., Moema, D., Molatlhegi, R., Ngobeni, P., Distribution of vanadium (V) species between soil and plants in the vicinity of vanadium mine. *Journal of hazardous materials* 2006, *137* (2), 649-653.

[31] ATSDR, "Toxicological profile for uranium." Atlanta, GA: U.S. Department of Health and Human Services, 2013.

[32] Zota, A. R., Willis, R., Jim, R., Norris, G. A., Shine, J. P., Duvall, R. M., Schaider, L. A., Spengler, J. D., Impact of mine waste on airborne respirable particulates in northeastern Oklahoma, United States. *Journal of the Air & Waste Management Association* 2009, *59* (11), 1347-1357.

[33] ATSDR, "Toxicological Profile for arsenic. Update." Atlanta, GA: U.S. Department of Health and Human Services, 2000.

[34] Dzik, E. A., Lobeck, H. L., Zhang, L., Burns, P. C., High-temperature calorimetric measurements of thermodynamic properties of uranyl arsenates of the meta-autunite group. *Chem. Geol.* 2018, *493*, 353-358.

[35] Coyte, R. M., Vengosh, A., Factors controlling the risks of co-occurrence of the redox-sensitive elements of arsenic, chromium, vanadium, and uranium in groundwater from the eastern United States. *Environ. Sci. Technol.* 2020, *54*, (7), 4367-4375.

[36] Das, N., Das, A., Sarma, K. P., Kumar, M., Provenance, prevalence and health perspective of co-occurrences of arsenic, fluoride and uranium in the aquifers of the Brahmaputra River floodplain. *Chemosphere* 2018, *194*, 755-772.

[37] Yadav, S. K., Ramanathan, A. L., Kumar, M., Chidambaram, S., Gautam, Y. P., Tiwari, C., Assessment of arsenic and uranium co-occurrences in groundwater of central Gangetic Plain, Uttar Pradesh, India. *Environ. Earth Sci.* 2020, *79*, (6), 154.

[38] Sobel, M., Sanchez, T. R., Zacher, T., Mailloux, B., Powers, M., Yracheta, J., Harvey, D., Best, L. G., Bear, A. B., Hasan, K., Thomas, E., Morgan, C., Aurand, D., Ristau, S., Olmedo, P., Chen, R., Rule, A., O'Leary, M., Navas-Acien, A., George, C. M., Bostick, B., Spatial relationship between well water arsenic and uranium in Northern Plains native lands. *Environ. Pollut.* 2021, *287*, 117655.

[39] Hoover, J., Gonzales, M., Shuey, C., Barney, Y., Lewis, J., Elevated arsenic and uranium concentrations in unregulated water sources on the Navajo Nation, USA. *Expos. Health.* 2017, *9*, (2), 113-124.

[40] Corkhill, C. L., Crean, D. E., Bailey, D. J., Makepeace, C., Stennett, M. C., Tappero, R., Grolimund, D., Hyatt, N. C., Multi-scale investigation of uranium attenuation by arsenic at an abandoned uranium mine, South Terras. *NPJ Mater. Degrad.* 2017, *1*, (1), 19.

[41] Dzik, E. A. L., Haylie L., Zhang, Lei; Burns, Peter C., Thermodynamic properties of phosphate members of the meta-autunite group: A high-temperature calorimetric study. *J. Chem. Thermodyn.* 2017, *114*, 165-171.

[42] Blake, J. M., Avasarala, S., Ali, A. M. S., Spilde, M., Lezama-Pacheco, J. S., Latta, D., Artyushkova, K., Ilgen, A. G., Shuey, C., Nez, C., Cerrato, J. M., Reactivity of As and U co-occurring in mine wastes in northeastern Arizona. *Chem. Geol.* 2019, *522*, 26-37.

[43] Hoover, J. H., Coker, E., Barney, Y., Shuey, C., Lewis, J., Spatial clustering of metal and metalloid mixtures in unregulated water sources on the Navajo Nation – Arizona, New Mexico, and Utah, USA. *Sci. Total Environ.* 2018, *633*, 1667-1678.

[44] Brockgreitens, J. W., Heidari, F., Abbas, A., Versatile process for the preparation of nanocomposite sorbents: phosphorus and arsenic removal. *Environ. Sci. Technol.* 2020, *54*, (14), 9034-9043.

[45] Gibbs, H. L. "Extraction of values from ores." US Patent US1999807A, April 30, 1935.

[46] McLeod, C., Uranium Mines and Mills May Have Caused Birth Defects among Navajo Indians. *Energy Resources* 1985, (12:49), 196-204.

[47] Churchill, W., American Indian lands: The Native ethic amid resource development. *Environment: Science and Policy for Sustainable Development* 1986, *28* (6), 13-34.

[48] Gofman, J. W., *Radiation and Human Health*. Random House, Inc. 1981.

[49] World Health Organization, "Uranium in drinking-water. Background document for preparation of WHO Guidelines for drinking-water quality." Geneva. World Health Organization (WHO/SDE/WSH/03.04/118) 2003.

[50] Domingo, J. L., Reproductive and developmental toxicity of natural and depleted uranium: a review. *Reproductive Toxicology* 2001, *15* (6), 603-609.

[51] Hund, L., Bedrick, E. J., Miller, C., Huerta, G., Nez, T., Ramone, S., Shuey, C., Cajero, M., Lewis, J., A Bayesian framework for estimating disease risk due to exposure to uranium mine and mill waste on the Navajo Nation. *Journal of the Royal Statistical Society: Series A (Statistics in Society)* 2015, *178* (4), 1069-1091.

[52] Santos, E. A., Ladeira, A. C., Recovery of uranium from mine waste by leaching with carbonate-based reagents. *Environmental science & technology* 2011, *45* (8), 3591-3597.

[53] Benjamin, M. M., *Water Chemistry*. Waveland Press, 2014.

[54] Ledin, M., Pedersen, K., The environmental impact of mine wastes—roles of microorganisms and their significance in treatment of mine wastes. *Earth-Science Reviews* 1996, *41* (1), 67-108.

[55] Chenoweth, W. L., "The Geology and Production Histroy of the Uranium Deposits in the Toreve A formation, Black Mesa, Apache county, Arizona." Arizona Geological Survey Contributed Report CR-90-A, 1990.

[56] Franczyk, K. J., "Stratigraphic revision and depositional environments of the Upper Cretaceous Toreva Formation in the northern Black Mesa area, Navajo and Apache Counties, Arizona." U.S. Geological Survey: Reston, VA 1988.

[57] Hostetler, P., Garrels, R., Transportation and precipitation of uranium and vanadium at low temperatures, with special reference to sandstone-type uranium deposits. *Economic Geology* 1962, *57* (2), 137-167.

[58] Mann, A. "Chemical ore genesis models for the precipitation of carnotite in calcrete;" Commonwealth Scientific and Industrial Research Organisation: 1974.

[59] Hillebrand, W., Carnotite and tyuyamunite and their ores in Colorado and Utah. *American Journal of Science* 1924, (45), 201-216.

[60] Crook, T., Blake, G., On Carnotite and an associated mineral complex from South Australia. *Mineralogical Magazine* 1910, *15* (71), 271-284.

[61] Sharma, R. K., Putirka, K. D., Stone, J. J., Stream sediment geochemistry of the upper Cheyenne River watershed within the abandoned uranium mining region of the southern Black Hills, South Dakota, USA. *Environmental Earth Sciences* 2016, *75* (9), 1-12.

[62] Xu, J., Zhu, S. Y., Luo, T. Y., Zhou, W., Li, Y. L., Uranium mineralization and its radioactive decay-induced carbonization in a black shale-hosted polymetallic sulfide ore layer, southwest China. *Economic Geology* 2015, *110* (6), 1643-1652.

[63] Abed, A. M., Saffarini, G. A., Sadaqah, R. M., Spatial distribution of uranium and vanadium in the upper phosphorite member in Eshidiyya basin, southern Jordan. *Arabian Journal of Geosciences* 2014, *7* (1), 253-271.

[64] Kim, C., Chi, C., Miller, S., Rosales, R., Sugihara, E., Akau, J., Rytuba, J., Webb, S., (Micro) spectroscopic analyses of particle size dependence on arsenic distribution and speciation in mine wastes. *Environmental science & technology* 2013, *47* (15), 8164-8171.

[65] Nelkin, D., Native Americans and nuclear power. *Science, Technology & Human Values* 1981, *6* (2), 2-13.

[66] USEPA, Jackpile-Paguate Uranium Mine NPL Listing Support Document, U.S. Government Publishing Office. *Washington, D.C.* 2013.

[67] Moench, R. H., Schlee, J. S., "Geology and uranium deposits of the Laguna district, New Mexico." US Government Printing Office: 1967.

[68] Risser, D. W., Davis, P. A., Baldwin, J. A., McAda, D. P., Aquifer tests at the Jackpile-Paguate uranium mine, Pueblo of Laguna, west-central New Mexico. *US Geol. Surv. Water Resour. Invest. Rep* 1984, *84* (4255), 26.

[69] Maxwell, C. H., Mesozoic stratigraphy of the Laguna-Grants region, in *Albuquerque Country II.*, NM Geol. Soc., 33rd Annual Field Conf, 1982; pp 261-266.

[70] Stumm, W., Morgan, J. J., *Aquatic chemistry: chemical equilibria and rates in natural waters*. John Wiley & Sons: 2012; Vol. 126.

[71] Langmuir, D., *Aqueous Environmental Geochemistry*, Prentice Hall, Upper Saddle River, NJ, 1997.

[72] Maurice, P. A., *Environmental surfaces and interfaces from the nanoscale to the global scale*. Wiley: 2009.

[73] Naeem, A., Westerhoff, P., Mustafa, S., Vanadium removal by metal (hydr) oxide adsorbents. *Water research* 2007, *41* (7), 1596-1602.

[74] Gerke, T. L., Scheckel, K. G., Schock, M. R., Identification and distribution of vanadinite (Pb5 (V^{5+} O_4) 3Cl) in lead pipe corrosion by-products. *Environmental science & technology* 2009, *43* (12), 4412-4418.

[75] Dong, W., Ball, W. P., Liu, C., Wang, Z., Stone, A. T., Bai, J., Zachara, J. M., Influence of calcite and dissolved calcium on uranium (VI) sorption to a Hanford subsurface sediment. *Environmental science & technology* 2005, *39* (20), 7949-7955.

[76] Merritt, R. *The Extractive Metallurgy of Uranium*, Colorado School of Mines Research Institute, 1971.

[77] Stewart, B. D., Nico, P. S., Fendorf, S., Stability of uranium incorporated into Fe (hydr) oxides under fluctuating redox conditions. *Environmental Science & Technology* 2009, *43* (13), 4922-4927.

[78] Stewart, B. D., Cismasu, A. C., Williams, K. H., Peyton, B. M., Nico, P. S., Reactivity of uranium and ferrous iron with natural iron oxyhydroxides. *Environmental science & technology* 2015, *49* (17), 10357-10365.

[79] Hsi, C. k. D., Langmuir, D., Adsorption of uranyl onto ferric oxyhydroxides: application of the surface complexation site-binding model. *Geochimica et Cosmochimica Acta* 1985, *49* (9), 1931-1941.

[80] Hattori, T., Saito, T., Ishida, K., Scheinost, A. C., Tsuneda, T., Nagasaki, S., Tanaka, S., The structure of monomeric and dimeric uranyl adsorption complexes on gibbsite: A combined DFT and EXAFS study. *Geochimica et Cosmochimica Acta* 2009, *73* (20), 5975-5988.

[81] Sachs, S., Bernhard, G., Sorption of U (VI) onto an artificial humic substance-kaolinite-associate. *Chemosphere* 2008, *72* (10), 1441-1447.

[82] Wang, Y., Frutschi, M., Suvorova, E., Phrommavanh, V., Descostes, M., Osman, A. A., Geipel, G., Bernier-Latmani, R., Mobile uranium (IV)-bearing colloids in a mining-impacted wetland. *Nature communications* 2013, *4*, 2942.

[83] Yang, Y., Saiers, J. E., Barnett, M. O., Impact of interactions between natural organic matter and metal oxides on the desorption kinetics of uranium from heterogeneous colloidal suspensions. *Environmental science & technology* 2013, *47* (6), 2661-2669.

[84] Luo, W., Gu, B., Dissolution and mobilization of uranium in a reduced sediment by natural humic substances under anaerobic conditions. *Environmental science & technology* 2008, *43* (1), 152-156.

[85] Mibus, J., Sachs, S., Pfingsten, W., Nebelung, C., Bernhard, G., Migration of uranium (IV)/(VI) in the presence of humic acids in quartz sand: A laboratory column study. *Journal of contaminant hydrology* 2007, *89* (3), 199-217.

[86] Steelink, C., Peer reviewed: investigating humic acids in soils. *Anal. Chem.* 2002, 74, 11, 326 A–333 A 2002.

[87] Keiluweit, M., Kleber, M., Molecular-level interactions in soils and sediments: the role of aromatic π-systems. *Environmental science & technology* 2009, *43* (10), 3421-3429.

[88] Yang, Y., Saiers, J. E., Xu, N., Minasian, S. G., Tyliszczak, T., Kozimor, S. A., Shuh, D. K., Barnett, M. O., Impact of natural organic matter on uranium transport through saturated geologic materials: from molecular to column scale. *Environmental science & technology* 2012, *46* (11), 5931-5938.

[89] McCarthy, J. F., Zachara, J. M., Subsurface transport of contaminants. *Environmental science & technology* 1989, *23* (5), 496-502.

[90] Kretzschmar, R., Borkovec, M., Grolimund, D., Elimelech, M., Mobile subsurface colloids and their role in contaminant transport. *Advances in agronomy* 1999, *66*, 121-193.

[91] Bone, S. E., Dynes, J. J., Cliff, J., Bargar, J. R., Uranium (IV) adsorption by natural organic matter in anoxic sediments. *Proceedings of the National Academy of Sciences* 2017, 201611918.

[92] Bhattacharyya, A., Campbell, K. M., Kelly, S. D., Roebbert, Y., Weyer, S., Bernier-Latmani, R., Borch, T., Biogenic non-crystalline U (IV) revealed as major component in uranium ore deposits. *Nature Communications* 2017, *8*.

[93] Mikutta, C., Langner, P., Bargar, J. R., Kretzschmar, R., Tetra-and hexavalent uranium forms bidentate-mononuclear complexes with particulate organic matter in a naturally uranium-enriched peatland. *Environmental science & technology* 2016, *50* (19), 10465-10475.

[94] Bargar, J. R., Williams, K. H., Campbell, K. M., Long, P. E., Stubbs, J. E., Suvorova, E. I., Lezama-Pacheco, J. S., Alessi, D. S., Stylo, M., Webb, S. M., Uranium redox transition pathways in acetate-amended sediments. *Proceedings of the National Academy of Sciences* 2013, *110* (12), 4506-4511.

[95] Veeramani, H., Alessi, D. S., Suvorova, E. I., Lezama-Pacheco, J. S., Stubbs, J. E., Sharp, J. O., Dippon, U., Kappler, A., Bargar, J. R., Bernier-Latmani, R., Products of abiotic U (VI) reduction by biogenic magnetite and vivianite. *Geochimica et Cosmochimica Acta* 2011, *75* (9), 2512-2528.

[96] Lovley, D. R., Phillips, E. J., Microbial reduction of uranium. *Nature* 1991, *350* (6317), 413.

[97] Bargar, J. R., Bernier-Latmani, R., Giammar, D. E., Tebo, B. M., Biogenic uraninite nanoparticles and their importance for uranium remediation. *Elements* 2008, *4* (6), 407-412.

[98] O'Loughlin, E. J., Kelly, S. D., Cook, R. E., Csencsits, R., Kemner, K. M., Reduction of uranium (VI) by mixed iron (II)/iron (III) hydroxide (green rust): formation of UO2 nanoparticles. *Environmental science & technology* 2003, *37* (4), 721-727.

[99] Nenoff, T. M., Jacobs, B. W., Robinson, D. B., Provencio, P. P., Huang, J., Ferreira, S., Hanson, D. J., Synthesis and low temperature in situ sintering of uranium oxide nanoparticles. *Chemistry of Materials* 2011, *23* (23), 5185-5190.

[100] Mkandawire, M., Dudel, E. G., Natural occurring uranium nanoparticles and the implication in bioremediation of surface mine waters. *Uranium, Mining and Hydrogeology* 2008, 487-496.

[101] Schindler, M., Lussier, A. J., Bellrose, J., Rouvimov, S., Burns, P. C., Kyser, T. K., Mobilization and agglomeration of uraninite nanoparticles: A nano-mineralogical

study of samples from the Matoush Uranium ore deposit. *American Mineralogist* 2017, *102* (9), 1776-1787.

[102] Singh, A., Catalano, J. G., Ulrich, K. U., Giammar, D. E., Molecular-scale structure of uranium (VI) immobilized with goethite and phosphate. *Environmental science & technology* 2012, *46* (12), 6594-6603.

[103] Troyer, L. D., Maillot, F., Wang, Z., Wang, Z., Mehta, V. S., Giammar, D. E., Catalano, J. G., Effect of phosphate on U (VI) sorption to montmorillonite: Ternary complexation and precipitation barriers. *Geochimica et Cosmochimica Acta* 2016, *175*, 86-99.

[104] Tokunaga, T. K., Kim, Y., Wan, J., Potential remediation approach for uranium-contaminated groundwaters through potassium uranyl vanadate precipitation. *Environmental science & technology* 2009, *43* (14), 5467-5471.

[105] Giammar, D. E., Hering, J. G., Time scales for sorption− desorption and surface precipitation of uranyl on goethite. *Environmental science & technology* 2001, *35* (16), 3332-3337.

[106] Singer, D. M., Zachara, J. M., Brown Jr, G. E., Uranium speciation as a function of depth in contaminated Hanford sediments-A micro-XRF, micro-XRD, and micro-and bulk-XAFS study. *Environmental science & technology* 2009, *43* (3), 630-636.

[107] Buck, E. C., Brown, N. R., Dietz, N. L., Contaminant uranium phases and leaching at the Fernald site in Ohio. *Environmental science & technology* 1995, *30* (1), 81-88.

[108] Kanematsu, M., Perdrial, N., Um, W., Chorover, J., O'Day, P. A., Influence of phosphate and silica on U (VI) precipitation from acidic and neutralized wastewaters. *Environmental science & technology* 2014, *48* (11), 6097-6106.

[109] Avasarala, S., Lichtner, P. C., Ali, A. M. S., González-Pinzón, R., Blake, J. M., Cerrato, J. M., Reactive Transport of U and V from Abandoned Uranium Mine Wastes. *Environmental Science & Technology* 2017, *51*, 12385-12393.

[110] Minelli, L., Veschetti, E., Giammanco, S., Mancini, G., Ottaviani, M., Vanadium in Italian waters: monitoring and speciation of V (IV) and V (V). *Microchemical journal* 2000, *67* (1), 83-90.

[111] Schock, M., Kelly, K. In *Vanadium chemistry essentials for treatment studies*, USEPA Inorganics Workshop, 2002.

[112] Veschetti, E., Maresca, D., Lucentini, L., Ferretti, E., Citti, G., Ottaviani, M., Monitoring of V (IV) and V (V) in Etnean drinking-water distribution systems by solid phase extraction and electrothermal atomic absorption spectrometry. *Microchemical journal* 2007, *85* (1), 80-87.

[113] Wehrli, B., Stumm, W., Vanadyl in natural waters: adsorption and hydrolysis promote oxygenation. *Geochimica et Cosmochimica Acta* 1989, *53* (1), 69-77.

[114] Baes Jr, C., Mesmer, R., *The Hydrolysis of Cations*. Wiley, New York: 1976.

[115] Burke, I. T., Mayes, W. M., Peacock, C. L., Brown, A. P., Jarvis, A. P., Gruiz, K., Speciation of arsenic, chromium, and vanadium in red mud samples from the Ajka spill site, Hungary. *Environmental science & technology* 2012, *46* (6), 3085-3092.

[116] Catana, G., Rao, R. R., Weckhuysen, B. M., Van Der Voort, P., Vansant, E., Schoonheydt, R. A., Supported vanadium oxide catalysts: Quantitative

spectroscopy, preferential adsorption of $V^{4+/5+}$, and Al_2O_3 coating of zeolite Y. *The Journal of Physical Chemistry B* 1998, *102* (41), 8005-8012.

[117] Manohar, D., Noeline, B., Anirudhan, T., Removal of vanadium (IV) from aqueous solutions by adsorption process with aluminum-pillared bentonite. *Industrial & engineering chemistry research* 2005, *44* (17), 6676-6684.

[118] Troyer, L. D., Stone, J. J., Borch, T., Effect of biogeochemical redox processes on the fate and transport of As and U at an abandoned uranium mine site: an X-ray absorption spectroscopy study. *Environmental Chemistry* 2014, *11* (1), 18-27.

[119] Curtis, G. P., Davis, J. A., Naftz, D. L., Simulation of reactive transport of uranium (VI) in groundwater with variable chemical conditions. *Water Resources Research* 2006, *42* (4).

[120] Hammond, G. E., Lichtner, P. C., Field-scale model for the natural attenuation of uranium at the Hanford 300 Area using high-performance computing. *Water Resources Research* 2010, *46* (9).

[121] Langmuir, D., *Aqueous Environmental Geochemistry*. Prentice Hall, 1997.

[122] Avasarala, S., Brearley, A. J., Spilde, M., Peterson, E., Jiang, Y. B., Benavidez, A., Cerrato, J. M. Crystal Chemistry of Carnotite in Abandoned Mine Wastes. *Minerals* 2020, *10*, 883.

[123] Chenoweth, W., Vanadium mining in the Carrizo Mountains, 1942-1947. San Juan County, New Mexico and Apache County Arizona: New Mexico Bureau of Mines and Mineral Resources, Open-file Report 1991, *378*, 33.

[124] Kelley, V. C., "Geology and technology of the Grants uranium region." State Bureau of Mines and Mineral Resources: Society of Economic Geologists, 1963.

[125] Nebbioso, A., Piccolo, A., Molecular characterization of dissolved organic matter (DOM): a critical review. *Anal Bioanal Chem* 2013, *405*, (1), 109-24.

[126] Potter, B. B., Wimsatt, J. C., "Method 415.3 - Measurement of total organic carbon, dissolved organic carbon and specific UV absorbance at 254 nm in source water and drinking water." U.S. Environmental Protection Agency, Washington, DC, 2005.

[127] Li, Y., Harir, M., Lucio, M., Kanawati, B., Smirnov, K., Flerus, R., Koch, B. P., Schmitt-Kopplin, P., Hertkorn, N., Proposed Guidelines for Solid Phase Extraction of Suwannee River Dissolved Organic Matter. *Analytical Chemistry* 2016, 88, (13), 6680-6688.

[128] Spirakis, C. S., The roles of organic matter in the formation of uranium deposits in sedimentary rocks. *Ore Geol. Rev.* 1996, 11, (1), 53-69.

[129] Cumberland, S. A., Douglas, G., Grice, K., Moreau, J. W., Uranium mobility in organic matter-rich sediments: A review of geological and geochemical processes. *Earth Sci. Rev.* 2016, 159, 160-185.

[130] Birdseye, H. S., Uranium deposits in northern Arizona. In *Black Mesa Basin (Northeastern Arizona)*, Anderson, R. Y., Harshbarger, J. W., New Mexico Geological Society 9th Annual Fall Field Conference Guidebook, 1958; pp 161-163.

[131] Deditius, A. P., Utsunomiya, S., Ewing, R. C., The chemical stability of coffinite, USiO4. nH2O; 0 <n< 2, associated with organic matter: a case study from Grants uranium region, New Mexico, USA. *Chem. Geol.* 2008, 1-4, (251), 33-49.

[132] Tsang, D., Chapter 2 - Influence of Natural Organic Matter on Contaminant Removal by Permeable Reactive Barrier. In *The Role of Colloidal Systems in Environmental Protection*, Fanun, M., Ed. Elsevier: Amsterdam, 2014; pp 19-40.

[133] Sparks, D. L., *Soil Physical Chemistry*, 2nd ed. New York : CRC Press, 1999.

[134] Sutton, R. S., Molecular structure in soil humic substances: the new view. *Environ. Sci. Technol.* 2005, (39), 9009–9015.

[135] Tipping, E., *Cation binding by humic substances*. Cambridge University Press, London. 2002.

[136] Darko, B., Jiang, J. Q., Kim, H., Machala, L., Zboril, R., Sharma, V. K., 8 - Advances Made in Understanding the Interaction of Ferrate(VI) with Natural Organic Matter in Water. In *Water Reclamation and Sustainability*, Ahuja, S., Ed. Elsevier: Boston, 2014; pp 183-197.

[137] de Melo, B., Motta, F. L., Santana, M. H. A., Humic acids: Structural properties and multiple functionalities for novel technological developments. *Materials Science and Engineering: C* 2016, *62*, 967-974.

[138] Averett, R. C., Leenheer, J., McKnight, D., Thorn, K. *Humic substances in the Suwannee River, Georgia; Interactions, properties, and proposed structures*, US Government Printing Office, 1994.

[139] Green, N., McInnis, D., Hertkorn, N., Maurice, P., Perdue, E., Suwannee River Natural Organic Matter: Isolation of the 2R101N Reference Sample by Reverse Osmosis. *Environ. Eng. Sci.* 2015, *32*, 38-44.

[140] El Hayek, E., Torres, C., Rodriguez-Freire, L., Blake, J. M., De Vore, C. L., Brearley, A. J., Spilde, M. N., Cabaniss, S., Ali, A. M. S., Cerrato, J. M., Effect of Calcium on the Bioavailability of Dissolved Uranium(VI) in Plant Roots under Circumneutral pH. *Environ. Sci. Technol.* 2018, *52*, 13089-13098.

[141] Lu, G., Haes, A. J., Forbes, T. Z., Detection and identification of solids, surfaces, and solutions of uranium using vibrational spectroscopy. *Coord. Chem. Rev.* 2018, *374*, 314-344.

[142] Yang, J. B., Volesky, B., Modeling uranium-proton ion exchange in biosorption. *Environmental Science & Technology* 1999, *33*, 4079-4085.

[143] Kim, K. W., Kim, Y. H., Lee, S. Y., Lee, J. W., Joe, K. S., Lee, E. H., Kim, J. S., Song, K., Song, K. C., Precipitation Characteristics of Uranyl Ions at Different pHs Depending on the Presence of Carbonate Ions and Hydrogen Peroxide. *Environmental Science & Technology* 2009, *43*, 2355-2361.

[144] Trenfield, M. A., Ng, J. C., Noller, B., Markich, S. J., van Dam, R. A., Dissolved organic carbon reduces uranium toxicity to the unicellular eukaryote Euglena gracilis. *Ecotoxicology* 2012, *21*, 1013-1023.

[145] Lavoie, M., Sabatier, S., Garnier-Laplace, J., Fortin, C., Uranium accumulation and toxicity in the green alga chlamydomonas reinhardtii is modulated by pH. *Environ. Toxicol. Chem.* 2014, *33*, 1372-1379.

[146] Granger H. C, S. E. S., Dean B. G., And Moore F. B., Sandstone-type uranium deposits at Ambrosia Lake, New Mexico; an interim report. *Econ. Geol.* 1961, *56*, 1179-1210.

[147] Ellsworth, H. V., Thucholite and uraninite from the Wallingford mine near Buckingham, Quebec. *Am. Mineral.* 1928, *13*, 442–448.

[148] Cumberland, S. A., Etschmann, B., Brugger, J., Douglas, G., Evans, K., Fisher, L., Kappen, P., Moreau, J. W., Characterization of uranium redox state in organic-rich Eocene sediments. *Chemosphere* 2018, *194*, 602-613.

[149] Regenspurg, S., Margot-Roquier, C., Harfouche, M., Froidevaux, P., Steinmann, P., Junier, P., Bernier-Latmani, R., Speciation of naturally-accumulated uranium in an organic-rich soil of an alpine region (Switzerland). *Geochimica et Cosmochimica Acta* 2010, *74*, 2082-2098.

[150] Adams, S. S., H. S. C., P. L. Hafen, H. Salek-Nejad, Interpretation of postdepositional processes related to the formation and destruction of the Jackpile-Paguate uranium deposit, Northwest New Mexico. *Econ. Geol.* 1978, *73*, 1635-1654.

[151] Volkov, I. V., Polyakov, E. V., Interaction of Humic Acids with Microelements/Radionuclides in Sorption Systems. *Radiochemistry* 2020, *62*, 141-160.

[152] Velasco, C. A., Artyushkova, K., Ali, A. M. S., Osburn, C. L., Gonzalez-Estrella, J., Lezama-Pacheco, J. S., Cabaniss, S. E., Cerrato, J. M., Organic Functional Group Chemistry in Mineralized Deposits Containing U(IV) and U(VI) from the Jackpile Mine in New Mexico. *Environmental Science & Technology* 2019, *53*, (10), 5758-5767

[153] Deshmukh, A. P., Pacheco, C., Hay, M. B., Myneni, S. C. B., Structural environments of carboxyl groups in natural organic molecules from terrestrial systems. Part 2: 2D NMR spectroscopy. *Geochim. Cosmochim. Ac.* 2007, *71*, (14), 3533-3544.

[154] Hay, M. B., Myneni, S. C. B., Structural environments of carboxyl groups in natural organic molecules from terrestrial systems. Part 1: Infrared spectroscopy. *Geochim. Cosmochim. Ac.* 2007, *71*, (14), 3518-3532.

[155] Mikutta, C., Kretzschmar, R., Spectroscopic evidence for ternary complex formation between arsenate and ferric iron complexes of humic substances. *Environ. Sci. Technol.* 2011, *45*, (22), 9550-9557.

[156] Stetten, L., Blanchart, P., Mangeret, A., Lefebvre, P., Le Pape, P., Brest, J., Merrot, P., Julien, A., Proux, O., Webb, S. M., Bargar, J. R., Cazala, C., Morin, G., Redox fluctuations and organic complexation govern uranium redistribution from U(IV)-phosphate minerals in a mining-polluted wetland soil, Brittany, France. *Environ. Sci. Technol.* 2018, *52*, 13099-13109.

[157] Avasarala, S. *Physical and chemical interactions affecting U and V transport from mine wastes*. Dissertation, University of New Mexico, 2018.

[158] Groudev, S., Georgiev, P., Spasova, I., Nicolova, M., Bioremediation of acid mine drainage in a uranium deposit. *Hydrometallurgy* 2008, *94*, (1-4), 93-99.

[159] Velasco, C. A., Brearley, A. J., Gonzalez-Estrella, J., Ali, A. M. S., Meza, M. I., Cabaniss, S. E., Thomson, B. M., Forbes, T. Z., Lezama Pacheco, J. S., Cerrato, J. M. From Adsorption to Precipitation of U(VI): What is the Role of pH and Natural Organic Matter? *Environ. Sci. Technol.* 2021, *55*, 16246-16256.

[160] Burdige, D. J., Kline, S. W., Chen, W., Fluorescent dissolved organic matter in marine sediment pore waters. *Marine Chemistry* 2004, *89*, (1), 289-311.

[161] Toet, S., Van Logtestijn, R. S. P., Kampf, R., Schreijer, M., Verhoeven, J. T. A., The effect of hydraulic retention time on the removal of pollutants from sewage treatment plant effluent in a surface-flow wetland system. *Wetlands* 2005, *25*, (2), 375-391.

[162] Avasarala, S., Torres, C., Ali, A. M. S., Thomson, B. M., Spilde, M. N., Peterson, E. J., Artyushkova, K., Dobrica, E., Lezama-Pacheco, J. S., Cerrato, J. M., Effect of bicarbonate and oxidizing conditions on U(IV) and U(VI) reactivity in mineralized deposits of New Mexico. *Chemical Geology* 2019, *524*, 345-355.

[163] Rakovan, J., Reeder, R. J., Elzinga, E. J., Cherniak, D. J., Tait, C. D., Morris, D. E., Structural Characterization of U(VI) in Apatite by X-ray Absorption Spectroscopy. *Environ. Sci. Technol*. 2002, *36*, (14), 3114-3117.

[164] Pan, Z., Giammar, D. E., Mehta, V., Troyer, L. D., Catalano, J. G., Wang, Z., Phosphate-Induced Immobilization of Uranium in Hanford Sediments. *Environ. Sci. Technol.* 2016, *50*, (24), 13486-13494.

[165] Wen Hang, H., Pan, Z., Giammar, D., Li, L., Enhanced uranium immobilization by phosphate amendment under variable geochemical and flow conditions: insights from reactive transport modeling. *Environ. Sci. Technol*. 2018, *52*, (10), 5841-5850.

[166] Wufuer, R., Wei, Y., Lin, Q., Wang, H., Song, W., Liu, W., Zhang, D., Pan, X., Gadd, G. M., Chapter Four - Uranium Bioreduction and Biomineralization. In *Adv. Appl. Microbiol*, Sariaslani, S., Gadd, G. M., Eds. Academic Press: 2017; Vol. 101, pp 137-168.

[167] Burns, P. C., Sigmon, G. E., *Uranium: cradle to grave*. Mineralogical Association of Canada Winnipeg: 2013; Vol. 43.

[168] Smedley, P. L., Kinniburgh, D. G., A review of the source, behaviour and distribution of arsenic in natural waters. *Appl Geochem* 2002, *17*, (5), 517-568.

[169] Crognale, S., Amalfitano, S., Casentini, B., Fazi, S., Petruccioli, M., Rossetti, S., Arsenic-related microorganisms in groundwater: a review on distribution, metabolic activities and potential use in arsenic removal processes. *Rev. Environ. Sci. Bio-Technol.* 2017, *16*, (4), 647-665.

[170] National Research Council, *Uranium mining in Virginia: scientific, technical, environmental, human health and safety, and regulatory aspects of uranium mining and processing in Virginia*. National Academies Press: 2012.

[171] Kapaj, S., Peterson, H., Liber, K., Bhattacharya, P., Human Health Effects From Chronic Arsenic Poisoning–A Review. *J. Environ. Sci, Part A* 2006, *41*, (10), 2399-2428.

[172] Finch, R., Murakami, T. "3. Systematics and Paragenesis of Uranium Minerals." In *Uranium: Mineralogy, Geochemistry, and the Environment,* edited by Peter C. Burns and Robert J. Finch, Berlin, Boston: De Gruyter, 1999, pp. 91-180.

[173] Troyer, L. D., Tang, Y., Borch, T., Simultaneous reduction of arsenic(v) and uranium(vi) by mackinawite: role of uranyl arsenate precipitate formation. *Environ. Sci. Technol.* 2014, *48*, (24), 14326-14334.

[174] Liu, H. K., Ramachandran, E., Chen, Y. H., Chang, W. J., Lii, K. H., High-temperature, high-pressure hydrothermal synthesis, characterization, and structural relationships of layered uranyl arsenates. *Inorg. Chem.* 2014, *53*, (17), 9065-9072.

[175] Chernorukov, N. G., Karyakin, N. V., The physical chemistry of the compounds MIP(As)UO_6 (MI=H, Li, Na, K, Rb, Cs) and their crystalline hydrates. *Russ. Chem. Rev.* 1995, *64*, (10), 913-927.

[176] Nipruk, O., Chernorukov, N., Pykhova, Y. P., Godovanova, N., Eremina, A., State of uranyl phosphates and arsenates in aqueous solutions. *Radiochemistry* 2011, *53*, (5), 483-490.

[177] Gorman-Lewis, D., Shvareva, T., Kubatko, K. A., Burns, P. C., Wellman, D. M., McNamara, B., Szymanowski, J. E. S., Navrotsky, A., Fein, J. B., Thermodynamic properties of autunite, uranyl hydrogen phosphate, and uranyl orthophosphate from solubility and calorimetric measurements. *Environ. Sci. Technol.* 2009, *43*, (19), 7416-7422.

[178] Nipruk, O. V., Chernorukov, N. G., Elipasheva, E. V., Klinshova, K. A., Bakhmetev, M. O., State of uranyl arsenates $M^IAsUO_6 \cdot nH_2O$ (M^I–H^+, Li^+, Na^+, K^+, Rb^+, Cs^+, NH4$^+$) in aqueous solution. *J. Radioanal. Nucl. Chem.* 2020, *324*, (1), 233-244.

[179] Gorman-Lewis, D., Burns, P. C., Fein, J. B., Review of uranyl mineral solubility measurements. *J. Chem. Thermodyn.* 2008, *40*, (3), 335-352.

[180] Gorman-Lewis, D., Fein, J. B., Burns, P. C., Szymanowski, J. E., Converse, J., Solubility measurements of the uranyl oxide hydrate phases metaschoepite, compreignacite, Na–compreignacite, becquerelite, and clarkeite. *J. Chem. Thermodyn.* 2008, *40*, (6), 980-990.

[181] Chernorukov, N., Nipruk, O., Knyazev, A., Pykhova, Y. P., Synthesis and study of uranyl arsenate $(UO_2)_3(AsO_4)_2 \cdot 12H_2O$. *Russ. J. Inorg. Chem.* 2011, *56*, (2), 163-167.

[182] Gezahegne, W. A., Hennig, C., Tsushima, S., Planer-Friedrich, B., Scheinost, A. C., Merkel, B. J., EXAFS and DFT investigations of uranyl arsenate complexes in aqueous solution. *Environ. Sci. Technol.* 2012, *46*, (4), 2228-2233.

[183] He, M., Liu, X., Cheng, J., Lu, X., Zhang, C., Wang, R., Uranyl arsenate complexes in aqueous solution: Insights from first-principles molecular dynamics simulations. *Inorg. Chem.* 2018, *57*, (10), 5801-5809.

[184] Rutsch, M., Geipel, G., Brendler, V., Bernhard, G., Nitsche, H., Interaction of uranium(VI) with arsenate(V) in aqueous solution studied by time-resolved laser-induced fluorescence spectroscopy (TRLFS). *Radiochim. Acta*, 1999, *86*, 135.

[185] Johnson, D. B., Hallberg, K. B., Acid mine drainage remediation options: a review. *Sci. Total Environ.* 2005, *338*, (1), 3-14.

[186] Santomartino, S., Webb, J. A., Estimating the longevity of limestone drains in treating acid mine drainage containing high concentrations of iron. *J. Appl. Geochem.* 2007, *22*, (11), 2344-2361.

[187] Aziz, H. A., Adlan, M. N., Ariffin, K. S., Heavy metals (Cd, Pb, Zn, Ni, Cu and Cr(III)) removal from water in Malaysia: Post treatment by high quality limestone. *Bioresour. Technol.* 2008, *99*, (6), 1578-1583.

[188] Lizama-Allende, K., Henry-Pinilla, D., Diaz-Droguett, D. E. J. W., Air,; Pollution, S., Removal of Arsenic and Iron from Acidic Water Using Zeolite and Limestone: Batch and Column Studies. *Water Air Soil Pollut.* 2017, *228*, (8), 275.

[189] Sdiri, A., Higashi, T., Jamoussi, F., Bouaziz, S., Effects of impurities on the removal of heavy metals by natural limestones in aqueous systems. *J. Environ. Manage.* 2012, *93*, (1), 245-253.

[190] Godelitsas, A., Astilleros, J. M., Hallam, K., Harissopoulos, S., Putnis, A., Interaction of Calcium Carbonates with Lead in Aqueous Solutions. *Environ. Sci. Technol.* 2003, *37*, (15), 3351-3360.

[191] Labastida, I., Armienta, M. A., Lara-Castro, R. H., Aguayo, A., Cruz, O., Ceniceros, N., Treatment of mining acidic leachates with indigenous limestone, Zimapan Mexico. *J. Hazard. Mater.* 2013, *262*, 1187-1195.

[192] Robbins, E. I., Cravotta, C. A., Savela, C. E., Nord, G. L., Hydrobiogeochemical interactions in 'anoxic' limestone drains for neutralization of acidic mine drainage. *Fuel* 1999, *78*, (2), 259-270.

[193] Gonzalez-Estrella, J., Meza, I., Burns, A. J., Ali, A. M. S., Lezama-Pacheco, J. S., Lichtner, P., Shaikh, N., Fendorf, S., Cerrato, J. M., Effect of bicarbonate, calcium, and pH on the reactivity of As(V) and U(VI) mixtures. *Environ. Sci. Technol.* 2020.

[194] DeVore, C., Freire, L. R., Gonzalez-Estrella, J., Villa, N., Ali, A. M., Ducheneaux, C., Artyushkova, K., Cerrato, J., Investigation of the effect of microbial processes on arsenic stability in sediments from Cheyenne River, South Dakota, USA. *Abstr. Pap. Am. Chem. Soc.* 2019, *257*, 1.

[195] Kaya, A., Yukselen, Y., Zeta potential of clay minerals and quartz contaminated by heavy metals. *Can. Geotech. J.* 2005, *42*, (5), 1280-1289.

[196] Alroudhan, A., Vinogradov, J., Jackson, M. D., Zeta potential of intact natural limestone: Impact of potential-determining ions Ca, Mg and SO4. *Colloids Surf. A: Physicochem. Eng. Asp.* 2016, *493*, 83-98.

[197] Jie, Z., Weiqing, W., Jing, L., Yang, H., Qiming, F., Hong, Z., Fe(III) as an activator for the flotation of spodumene, albite, and quartz minerals. *Miner. Eng.* 2014, *61*, 16-22.

[198] Locock, A. J., Burns, P. C., Duke, M. J. M., Flynn, T. M., Monovalent cations in structures of the meta-autunite group. *Can. Mineral.* 2004, *42*, (4), 973-996.

[199] Cox, J. D., Wagman, D. D., Medvedev, V. A., *CODATA Key Values for Thermodynamics*. 1989.

[200] Ondrus, P., Veselovsky, F., Skala, R., Cisarova, I., Hlousek, J., Fryda, J., Vavrin, I., Cejka, J., Gabasova, A., New naturally occurring phases of secondary origin from Jachymov (Joachimsthal). *J. Czech Geol. Soc.* 1997, *42*, (4), 77-108.

[201] Clavier, N., Crétaz, F., Szenknect, S., Mesbah, A., Poinssot, C., Descostes, M., Dacheux, N., Vibrational spectroscopy of synthetic analogues of ankoleite, chernikovite and intermediate solid solution. *Spectrochim. Acta B: At. Spectrosc.* 2016, *156*, 143-150.

[202] Ross, M., Evans Jr, H., Studies of the torbernite minerals (III): role of the interlayer oxonium, potassium, and ammonium ions, and water molecules. *Am. Mineral.* 1965, *50*, (1-2), 1-12.

[203] Ross, M., Evans Jr, H., Appleman, D., Studies of the torbernite minerals (II): The crystal structure of meta-torbernite. *Am. Mineral.* 1964, *49*, (11-12), 1603-1621.

[204] Ross, M., Evans Jr, H. T., Studies of the torbernite minerals (I): The crystal structure of abernathyite and the structurally related compounds

$NH_4(UO_2AsO_4)\cdot 3H_2O$ and $K(H_3O)(UO_2AsO_4)_2\cdot 6H_2O$. *Am. Mineral.* 1964, *49*, 1578-1602.

[205] Meza, I., Gonzalez-Estrella, J., Burns, P. C., Rodriguez, V., Velasco, C. A., Sigmon, G. E., Szymanowski, J. E. S., Forbes, T. Z., Applegate, L. M., Ali, A. M. S., Lichtner, P., Cerrato, J. M. Solubility and Thermodynamic Investigation of Meta-Autunite Group Uranyl Arsenate Solids with Monovalent Cations Na and K. *Environ Sci & Tech* 2023, *57*, 255-265.

[206] Morosin, B., Hydrogen uranyl phosphate tetrahydrate, a hydrogen ion solid electrolyte. *Acta Crystallographica - Section B: Structural Crystallography and Crystal Chemistry* 1978, *34*, (12), 3732.

[207] Ivanova, A. G., Budantseva, N. A., Fedoseev, A. M., Synthesis, crystal structure, and properties of new actinide(VI) arsenates $(H_3O)[(AnO_2)(AsO_4)]\cdot 3H_2O$ (An = U, Np, Pu). *Radiochemistry* 2017, *59*, (2), 119-123.

[208] Shilton, M. G., Howe, A. T., Rapid H^+ conductivity in hydrogen uranyl phosphate-A solid H^+ electrolyte. *Materials Research Bulletin* 1977, *12*, (7), 701-706.

[209] Benjamin, M. M., *Water Chemistry*. Boston: McGraw-Hill, 2002.

[210] Ulrich, K. U., Ilton, E. S., Veeramani, H., Sharp, J. O., Bernier-Latmani, R., Schofield, E. J., Bargar, J. R., Giammar, D. E., Comparative dissolution kinetics of biogenic and chemogenic uraninite under oxidizing conditions in the presence of carbonate. *Geochim. Cosmochim. Acta* 2009, *73*, (20), 6065-6083.

[211] Reinoso-Maset, E., Rates and mechanisms of uranyl oxyhydroxide mineral dissolution. *Geochim. Cosmochim. Acta* 2017, *207*, 298.

[212] Akcil, A., Koldas, S., Acid Mine Drainage (AMD): causes, treatment and case studies. *J. Clean. Prod.* 2006, *14*, (12), 1139-1145.

[213] Meza, I., Jemison, N., Gonzalez-Estrella, J., Burns, P. C., Rodriguez, V., Sigmon, G. E., Szymanowski, J. E. S., Ali, A. M. S., Gagnon, K., Cerrato, J. M., Lichtner, P. Kinetics of Na- and K- uranyl arsenate dissolution, *Chemical Geology*, 2023, *636*, 121642.

Puzzle 4

Spotted Uranium Killer

by rockratzero

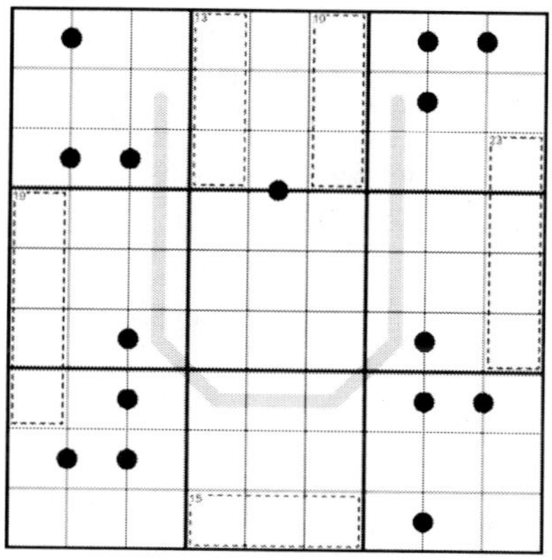

Rules

Place the digits 1-9 in each row, column, and 3x3 region without repeating.

Digits on the gray lines alternate between odd and even numbers.

Digits joined by black dots must be in a 2:1 ratio (one is double the other).

Digits contained within the dashed cages ("killer cages") must sum to the value in the top left corner of the cage.

Link to SudokuPad online: https://tinyurl.com/rrz-uranium4.

About the Editor

William A. Alexander
Department of Chemistry
The University of Memphis
Memphis, TN, USA

Dr. Alexander received a BS Chemistry degree from Marshall University, where he was a member of the Society of Yeager Scholars. He earned his PhD from Virginia Tech in 2009 under the tutelage of Dr. Diego Troya and Dr. John Morris. After postdoctoral work at Virginia Tech and Montana State University-Bozeman, Dr. Alexander was appointed to the faculty at The University of Memphis in 2012, where he is currently an Associate Professor of Chemistry.

Dr. Alexander's research interests focus on characterizing reactivity at the gas-surface interface using surface science and computational chemistry techniques. Dr. Alexander also researches aspects of water resource protection, where he is developing computational tools for contaminant property screening, as well as a program using aerial amphibious drones for sampling and analysis of surface waters in emergency operations. His work has been funded as a DRONES research fellow of the FedEx Institute of Technology, and by the Tennessee Department of Environment and Conservation, the National Science Foundation, and the US NNSA Office of Defense Nuclear Nonproliferation Research and Development program with Argonne National Lab.

Puzzle Solutions

Classic Uranium

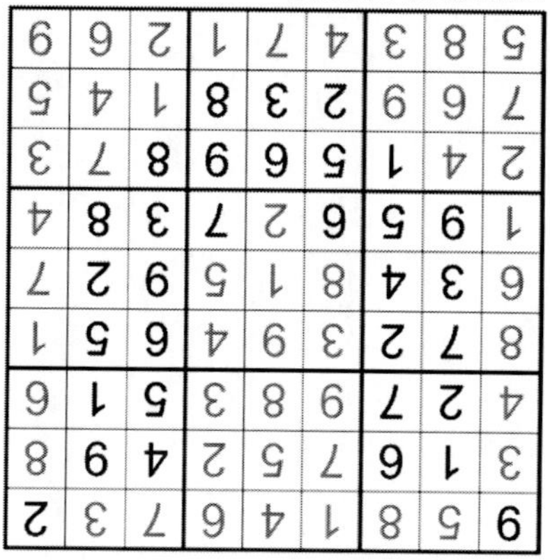

Uranium Hexafluoride Interactions

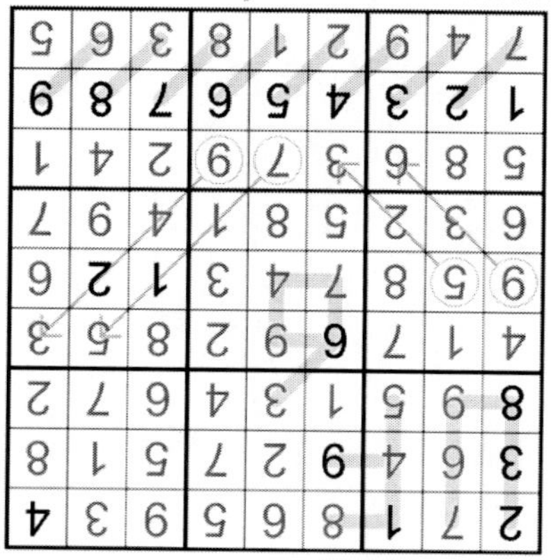

Uranium on the Table

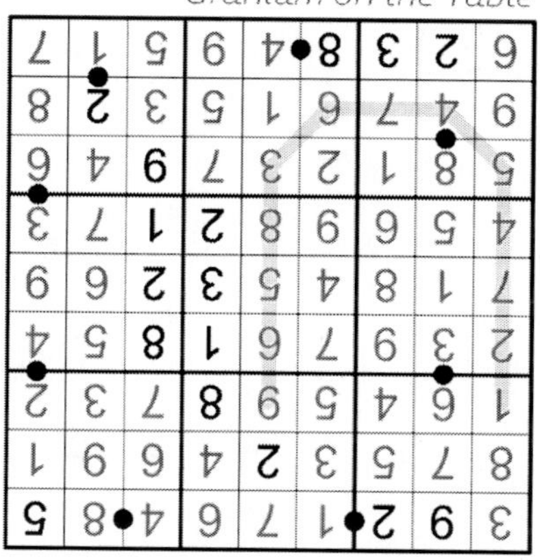

Spotted Uranium Killer

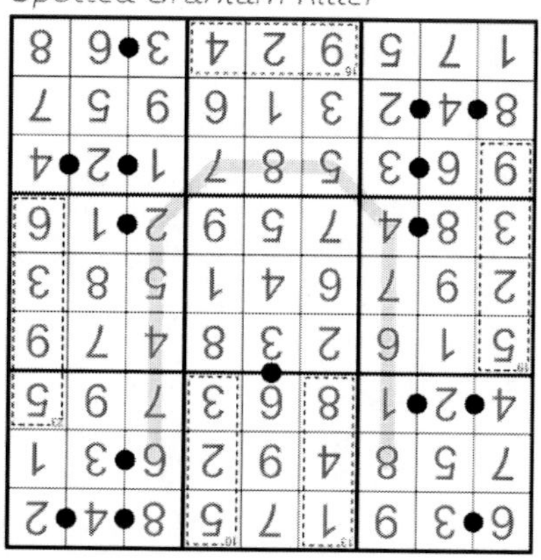

I hope you had fun with these puzzles!

Feel free to reach out via email rockratzero@gmail.com with any issues about the puzzles, or with any business/professional inquiries.

- rockratzero

Index

E

F

G

H

I

K

L

M

X

Y